The Role of True Finiteness in the Admissible Recursively Enumerable Degrees

MEMOIRS
of the
American Mathematical Society

Number 854

The Role of True Finiteness in the Admissible Recursively Enumerable Degrees

Noam Greenberg

May 2006 • Volume 181 • Number 854 (third of 5 numbers) • ISSN 0065-9266

American Mathematical Society
Providence, Rhode Island

2000 *Mathematics Subject Classification.*
Primary 03D60; Secondary 03D25, 03D30.

Library of Congress Cataloging-in-Publication Data

Greenberg, Noam, 1974–
The role of true finiteness in the admissible recursively enumerable degrees / Noam Greenberg.
p. cm. — (Memoirs of the American Mathematical Society, ISSN 0065-9266 ; no. 854)
"Volume 181, number 854 (third of 5 numbers)."
Includes bibliographical references.
ISBN 0-8218-3885-7 (alk. paper)
1. Recursion theory. 2. Recursively enumerable sets. 3. Finite, The. I. Title. II. Series.

QA3 .A57 no. 854
[QA9.6]
510 s—dc22
[511.3′5] 2006040677

Memoirs of the American Mathematical Society

This journal is devoted entirely to research in pure and applied mathematics.

Subscription information. The 2006 subscription begins with volume 179 and consists of six mailings, each containing one or more numbers. Subscription prices for 2006 are US$624 list, US$499 institutional member. A late charge of 10% of the subscription price will be imposed on orders received from nonmembers after January 1 of the subscription year. Subscribers outside the United States and India must pay a postage surcharge of US$31; subscribers in India must pay a postage surcharge of US$43. Expedited delivery to destinations in North America US$35; elsewhere US$130. Each number may be ordered separately; *please specify number* when ordering an individual number. For prices and titles of recently released numbers, see the New Publications sections of the *Notices of the American Mathematical Society.*

Back number information. For back issues see the *AMS Catalog of Publications.*

Subscriptions and orders should be addressed to the American Mathematical Society, P. O. Box 845904, Boston, MA 02284-5904, USA. *All orders must be accompanied by payment.* Other correspondence should be addressed to 201 Charles Street, Providence, RI 02904-2294, USA.

Memoirs of the American Mathematical Society is published bimonthly (each volume consisting usually of more than one number) by the American Mathematical Society at 201 Charles Street, Providence, RI 02904-2294, USA. Periodicals postage paid at Providence, RI. Postmaster: Send address changes to Memoirs, American Mathematical Society, 201 Charles Street, Providence, RI 02904-2294, USA.

This publication is indexed in *Science Citation Index®*, *SciSearch®*, *Research Alert®*, *CompuMath Citation Index®*, *Current Contents®/Physical, Chemical & Earth Sciences.*
Printed in the United States of America.

∞ The paper used in this book is acid-free and falls within the guidelines
established to ensure permanence and durability.
Visit the AMS home page at http://www.ams.org/

10 9 8 7 6 5 4 3 2 1 11 10 09 08 07 06

Contents

Abstract

When attempting to generalize recursion theory to admissible ordinals, it may seem as if all classical priority constructions can be lifted to any admissible ordinal satisfying a sufficiently strong fragment of the replacement scheme. We show, however, that this is not always the case. In fact, there are some constructions which make an essential use of the notion of finiteness which cannot be replaced by the generalized notion of α-finiteness. As examples we discuss both codings of models of arithmetic into the recursively enumerable degrees, and non-distributive lattice embeddings into these degrees. We show that if an admissible ordinal α is effectively close to ω (where this closeness can be measured by size or by cofinality) then such constructions may be performed in the α-r.e. degrees, but otherwise they fail. The results of these constructions can be expressed in the first-order language of partially ordered sets, and so these results also show that there are natural elementary differences between the structures of α-r.e. degrees for various classes of admissible ordinals α. Together with coding work which shows that for some α, the theory of the α-r.e. degrees is complicated, we get that for every admissible ordinal α, the α-r.e. degrees and the classical r.e. degrees are not elementarily equivalent.

Received by the editor 23rd of September, 2004.

1991 *Mathematics Subject Classification.* Primary 03D60; Secondary 03D25, 03D30.

Key words and phrases. True finiteness, admissible ordinal, recursively enumerable, lattice embedding.

The results in this work are contained in my doctoral dissertation, written at Cornell University under the guidance of Richard A. Shore, whom I would like to warmly thank. Partially supported by NSF Grant DMS-0100035.

CHAPTER 1

Introduction

The study of recursive ordinals and hyperarithmetic sets that began with the work of Church and Kleene [**CK37**], Church [**Chu38**] and Kleene [**Kle38**] suggested many analogies between the Π^1_1 and hyperarithmetic sets and the recursively enumerable and recursive ones, respectively. The analogy was not perfect, however. At the basic level, for example, the range of a hyperarithmetic function on a hyperarithmetic set is always hyperarithmetic rather than an arbitrary Π^1_1 set. At a deeper level, all nonhyperarithmetic Π^1_1 sets are of the same hyperarithmetic degree. Kreisel [**Kre61**] studied this situation and came to the realization that while Π^1_1 is analogous to r.e., the correct analog for hyperarithmetic is not recursive but finite. This insight lead first to the development by Kreisel and Sacks [**KS63, KS65**] of metarecursion theory as the study of recursion theory on the recursive ordinals (those less than ω_1^{CK}, the first nonrecursive ordinal) or, equivalently, on their notations in a Π^1_1 path through Kleene's $\mathcal{O}$. In this setting, the meta-r.e. subsets of ω are the Π^1_1 ones and the metafinite ones are hyperarithmetic.

Another approach to generalizing recursion theory to ordinals started with Takeuti's [**Tak60, Tak65**] development of Gödel's [**G39**] constructible universe L through a recursion theory on the class of all ordinals. These two approaches came together in the common generalization of recursion on admissible ordinals of Kripke [**Kri64**] and Platek [**Pla65**]. Here the domain of discourse is an ordinal α or the initial segment L_α of L up to α for admissible α, i.e. L_α satisfies Σ_1-replacement. In this vein, α-r.e. is Σ_1 over L_α, α-recursive is then Δ_1 over L_α while α-finite means a member of L_α. These notions coincide with those of metarecursion theory when $\alpha = \omega_1^{CK}$.

We should also note that care has to be taken in the definition of "α-recursive in", the analog of Turing reducibility. Here too, the crucial issue is that of finiteness. It no longer suffices to require that one be able to answer single membership question about A in a computation from B to say that A is reducible to B. Instead one defines α-reducible, $\leqslant_\alpha$, by requiring that all α-finite sets of such questions about A can be computed on the basis of α-finitely much information about B.

The motivation and goals for generalizing recursion theory in this way included the hopes of elucidating the underlying nature of the notions fundamental to recursion theory and the essences of the constructions that are used to prove its most important theorems. In accordance by Kreisel's insight, a prominent role should be played by the analysis of finiteness along with recursive and recursively enumerable. Such an analysis might lead to a good axiomatic treatment or reveal approaches that would be less dependent on the specific combinatorial properties of ω exploited in these notions and constructions. In this way the study might also produce applications to both classical recursion theory and other domains (set theory, model theory, proof theory and, in hindsight, computer science) where the notions of effectiveness play many roles.

It was relatively easy to formalize the basic notions of recursion theory in these settings but also in much more general ones. Kreisel's test of a generalization worthy of investigation was the Freidberg-Muchnik theorem solving Post's problem by showing that there are incomparable r.e. degrees. As Sacks [**Sac90**, p. ix] puts it, this brings us from the static or syntactic realm into the dynamic one. It is in this domain that priority arguments and the deeper investigations into the notion of enumerability and relative computability were developed in classical recursion theory. First metarecursion theory (Sacks [**Sac66**]) and then α-recursion theory (Sacks and Simpson [**SS72**]) passed this test.

The route to the solution to Post's problem in α-recursion theory was the ability to make Σ_1-replacement suffice for arguments that in classical recursion theory seemed to naturally rely on Σ_2-replacement (or induction). Further investigations in α-recursion theory indicated that many of the more complicated priority arguments of the classical subject used yet higher levels of replacement and did not generalize so readily to all admissible α. The density theorem was successfully generalized to all admissible α (Shore [**Sho76b**]) but to this day the theorems epitomizing the basic construction of classical recursion theory have not been settled for all admissible ordinals. Almost always more admissibility suffices and at times other conditions as well. Early examples include the existence of an incomplete high α-r.e. degree (Shore [**Sho76a**]) and minimal pairs (Lerman and Sacks [**LS72**]) for which Σ_2 admissibility suffices and at times something less. Eventually, an elementary difference between the r.e. degrees and the α-r.e. degrees for some α was established by finding certain admissible ordinals for which, contrary to Lachlan's [**Lac76**] nonsplitting theorem, one can combine splitting and density for all pairs of α-r.e. degrees (Shore [**Sho78**]). (That is, for certain α it is always possible to find, for every pair $\mathbf{a} < \mathbf{b}$ of α-r.e. degrees, two incomparable α-r.e. degrees $\mathbf{b}_0$ and $\mathbf{b}_1$ between $\mathbf{a}$ and $\mathbf{b}$ such that $\mathbf{b}_0 \vee \mathbf{b}_1 = \mathbf{b}$.) This work did indeed elucidate the role of various replacement or induction-like principles in recursion theoretic arguments and much later played a role in analyzing such arguments in reverse mathematics (e.g. Slaman and Woodin [**SW89**] and Mytilinaios [**Myt89**]). Other aspects of generalized recursion theory found applications in complexity theory (e.g. Shinoda and Slaman [**SS90**]). They did not however have much to say directly about the role of finiteness. Moreover, once the basic techniques are understood, all these constructions can be fairly easily carried out in metarecursion theory.

The crucial fact about ω_1^{CK} needed to carry out all these arguments is that there is a metarecursive projection of ω onto ω_1^{CK}. This allows one to arrange priority requirements in an ω list and so carry out constructions in such a way that one only ever really needs to worry about there being truly finitely many predecessors of any requirement. For example, density was proved by Driscoll [**Dri68**] and minimal pairs constructed by Sukonick [**Suk69**]. It seemed as if everything one could do in classical recursion theory could be done in metarecursion theory as well. It was in this setting that Sacks [**Sac63**] posed as his final question whether $\mathcal{R}_{\omega_1^{CK}}$, the meta-r.e. degrees with ω_1^{CK}-reducibility, and $\mathcal{R}$, the r.e. ones with Turing reducibility, are elementarily equivalent. This seemed possible at the time. Indeed, at that time people still thought that there should be some nice characterization of the structure $\mathcal{R}$ that would indicate that it was simple in some way. Shoenfield's conjecture that it was ω-saturated and so categorical had been disproven with the construction of a minimal pair of r.e. degrees but, nonetheless, Sacks still conjectured in [**Sac63**] that the theory was decidable and that the structure was isomorphic to the degrees r.e. in and above $\mathbf{d}$ for every degree $\mathbf{d}$.

Both of these conjectures turned out to be false (Harrington-Shelah [**HS82**], Shore [**Sho82**]). Indeed, these results and others showed that $\mathcal{R}$ was very complicated in various ways. Shore [**Sho82**] showed that it is not recursively presentable and later Harrington and Slaman and Slaman and Woodin (see Slaman [**Sla91**]) showed that its theory is recursively isomorphic to true arithmetic. These sorts of results changed the paradigm for understanding $\mathcal{R}$ from a hope for simplicity to an approach to its characterization by its complexity. (For more of the history and further discussion, see Shore [**Sho97**] and [**Sho99**]). Once one had this view of $\mathcal{R}$, it became natural to believe that the answer to Sacks' question was "no" just because it seemed that one could prove all the results of classical recursion theory in metarecursion theory. If the meta-r.e. degrees, like the r.e. ones, are as complicated as possible then $\mathcal{R}_{\omega_1^{CK}}$ is more complicated than $\mathcal{R}$. In this way, Odell [**Ode83**] established an analog of Shore [**Sho82**] for the meta-r.e. degrees to show that $\mathcal{R}_{\omega_1^{CK}}$ is not arithmetically presentable and so not isomorphic to $\mathcal{R}$. Once Harrington and Slaman and Slaman and Woodin had proven that the theory of $\mathcal{R}$ is recursively isomorphic to true arithmetic, it became "morally certain" that the two structures are not even elementarily equivalent.

Shore and Slaman, as announced in Shore [**Sho97**], managed to carry out enough of the relevant constructions in metarecursion theory to prove this result. The proof was fairly elaborate and required lifting several major theorems of classical recursion theory to ω_1^{CK}. It also failed to give a full characterization of the degree of the theory of $\mathcal{R}_{\omega_1^{CK}}$. The expected result was that it should be recursively isomorphic to the theory of $\langle L_{\omega_1^{CK}}, \in\rangle$ or, equivalently, of degree $\mathcal{O}^{(\omega)}$. This result awaited further developments in classical recursion theory. Nies, Shore and Slaman [**NSS98**] provided a definable standard model of arithmetic in $\mathcal{R}$ and so a more direct proof that the degree of its theory is $\mathbf{0}^{(\omega)}$. In [**GSS**], the same original intuition from the 60s about the similarity of $\mathcal{R}$ and $\mathcal{R}_{\omega_1^{CK}}$ was followed, to lift enough of Nies, Shore and Slaman [**NSS98**] to metarecursion theory to prove that a standard model of arithmetic with a predicate for $\mathcal{O}$ is definable in $\mathcal{R}_{\omega_1^{CK}}$ and so its theory, as expected, is recursively isomorphic to both that of $L_{\omega_1^{CK}}$ and to $\mathcal{O}^{(\omega)}$. These results thus answered Sacks's original question by providing an elementary difference between $\mathcal{R}$ and $\mathcal{R}_{\omega_1^{CK}}$. However, they did so by continuing along the path following the intuition that one can lift all constructions of r.e. degrees to ω_1^{CK} by using projectability to convert requirements lists to ones of length ω and to any admissible ordinal satisfying enough replacement to handle requirements in order type α.

These illusions are hereby dispelled in this present work, whose aim is to illuminate the role of true finiteness in various classical constructions in the setting of the r.e. *degrees* (Lerman and Simpson [**LS73**] and Lerman ([**Ler74**]) gave such results in the context of the lattice of r.e. sets). We show here that constructions given in [**NSS98**] can be performed in the α-r.e. degrees (for Σ_2-admissible α) if and only if the cofinality of α, as measured by some relatively effective class of functions, is ω. Another line of investigation considers construction which are used to embed some nondistributive lattices into the r.e. degrees. Lachlan ([**Lac72**]) has shown that the 1-3-1 lattice, the one of the two basic nondistributive lattices which includes a *critical triple*, is embeddable in the r.e. degrees. We show that this construction uses finiteness in an essential way; it can only be performed in the α-r.e. degrees (here α is any admissible ordinal) if α is countable in some effective sense. All of the work taken together shows that no $\mathcal{R}_\alpha$ is elementarily equivalent to $\mathcal{R}$.

Notation and Terminology. Rather than give a long list of definitions of the concepts used in this work and of the notation associate with them, we prefer to place each definition in its natural context as it comes up in the work. Many of these definitions make sense only together with facts concerning the objects involved; in the appendices we develop much of the theory which is needed to make most of the definitions comprehensible.

The basics of admissible recursion theory are developed in appendix A. It is there where we define our "playing ground", namely Jensen's J_α hierarchy, describe the notions of amenability and admissibility, discuss α-recursive enumerations and α-reductions, and in that context it makes sense to define nice functionals, which are not used in the standard texts of the field (Sacks [**Sac90**] and Chong [**Cho84**].) We also define there effective versions of cofinality, such as the Σ_n cofinality of α and the recursive cofinality of a degree.

Further notions of α-recursion theory include the jump operator (and the notion of lowness which accompanies it), described in appendix B, and the projectum ϱ^n_α, which is discussed in appendix C.

Notions from classical recursion theory, often more algebraic in nature (such as the lattices we try to embed in the degrees, or the coding of models of arithmetic in the degrees), are defined in the body of the text as they appear.

The notation used follows modern set-theoretic standards (see, for example, Jech [**Jec03**]). Thus for example we use $f``X$ to denote the pointwise image of X under the function f. $\subset$ means inclusion, whether proper or not (when we want to stress proper inclusion we will use $\subsetneq$). *Club* means closed and unbounded (usually in the fixed admissible ordinal α).

For recursion theory we use the functional notation which has become standard in recent years. The meaning of this notation in the context of admissible recursion theory is explained in appendix A. During effective constructions or enumerations we use Lachlan's notation ([**Lac79**]) of modifying objects and whole expressions by $[s]$ to denote they are viewed in stage s of the construction or enumeration. In general notation will be similar to the one used in [**Soa87**, XIV s.4].

1. The Results

In this thesis we show how the proximity of an admissible ordinal α to ω is reflected in the structure of the α-recursively enumerable degrees (which we denote by $\mathcal{R}_\alpha$).

1.1. Lattice Embeddings.

THEOREM 1.1. *There is a sentence ψ (in the language of partially ordered sets) such that for every admissible ordinal α, $\mathcal{R}_\alpha \models \psi$ iff $\varrho^2_\alpha = \omega$ and $\mathrm{cf}_{\Sigma_2(J_\alpha)}(\alpha) = \omega$.*

The sentence ψ states the existence of an embedding of the 1-3-1 lattice (also known as M_5; see [**Soa87**, IX 2.7] and figure 6.1) into $\mathcal{R}_\alpha$ with an incomplete top. The class of ordinals α such that $\mathcal{R}_\alpha \models \psi$ is exactly the class of admissible ordinals

α such that there is an incomplete degree $\mathbf{a} \in \mathcal{R}_\alpha$ such that $\operatorname{rcf}(\mathbf{a})$, the recursive cofinality of $\mathbf{a}$, is ω. For the definition of recursive cofinality and $\Sigma_2(J_\alpha)$ cofinality, see the end of appendix A. For the definition of the Σ_2-projectum ϱ^2_α see appendix C.

In chapter 5 we show (theorem 5.4) that if α is admissible and $\operatorname{rcf}(\mathbf{a}) > \omega$ then the 1-3-1 lattice cannot be embedded below $\mathbf{a}$; in fact, there is no weak critical triple below $\mathbf{a}$ (see definition in chapter 5). In chapter 6 (theorem 6.1) we show that if $\mathbf{a} \in \mathcal{R}_\alpha$ is collapsible then we can embed the 1-3-1 lattice below $\mathbf{a}$. In appendix D (where the notion of a collapsible degree is defined) we quote Shore ([**Sho76b**]) and see that if $\mathbf{a} \in \mathcal{R}_\alpha$ is incomplete and $\operatorname{rcf}(\mathbf{a}) = \omega$ then $\mathbf{a}$ is collapsible.

The following theorem is also an immediate consequence of these results:

THEOREM 1.2. *Let α be admissible. The following are equivalent for an incomplete degree $\mathbf{a} \in \mathcal{R}_\alpha$:*

(1) *There is a weak critical triple in $\mathcal{R}_\alpha(\leqslant \mathbf{a})$.*
(2) *There is an embedding of the 1-3-1 lattice into $\mathcal{R}_\alpha(\leqslant \mathbf{a})$.*
(3) $\operatorname{rcf}(\mathbf{a}) = \omega$.

This gives us two formulas φ_0 and φ_1 (which are not equivalent in the theory of partial orderings or in the theory of upper semi-lattices) such that for every admissible α,

$$\varphi_0(\mathcal{R}_\alpha) \;=\; \varphi_1(\mathcal{R}_\alpha) \;=\; \{\mathbf{a} \in \mathcal{R}_\alpha \,:\, \mathbf{a} < \mathbf{0}' \;\;\&\;\; \operatorname{rcf}(\mathbf{a}) = \omega\}.$$

1.2. Effective Successor Models. In chapter 3 we show that if α is Σ_2-admissible and $\operatorname{cf}_{\Sigma_3(J_\alpha)}(\alpha) = \omega$ then one can embed an effective successor copy of the standard model of arithmetic into $\mathcal{R}_\alpha$ below any promptly permitting degree $\mathbf{u}$, such that below $\mathbf{u}$ there is no least upper bound for the elements of the model. [See the definition of our coding of models of arithmetic in chapter 2 and that of an effective successor model in chapter 3. Prompt permission is discussed in appendix E.] In chapter 4 we show that for various classes of α, including the class of Σ_2-admissible ordinals such that $\operatorname{cf}_{\Sigma_3(J_\alpha)}(\alpha) > \omega$, there can be no such embedding below any low$_2$ degree $\mathbf{u}$.

We give some details. Let $\chi_{\texttt{effective}}(\bar{\mathbf{p}}, \bar{\mathbf{e}})$ be the formula stating that $\bar{\mathbf{p}}$ codes an effective successor model of arithmetic, witnessed by $\bar{\mathbf{e}}$; see chapter 3 (we denote the model coded by $\bar{\mathbf{p}}$ by $M_{\bar{\mathbf{p}}}$). Theorem 3.1 states that if α is Σ_2-admissible, $\operatorname{cf}_{\Sigma_3(J_\alpha)}(\alpha) = \omega$ and $\mathbf{u}$ is a promptly permitting degree then there are some $\bar{\mathbf{p}}, \bar{\mathbf{e}} \leqslant \mathbf{u}$ satisfying $\chi_{\texttt{effective}}$ such that $M_{\bar{\mathbf{p}}}$ is standard (i.e. is isomorphic to the standard model of arithmetic), and such that there is an exact pair $\mathbf{c}_0, \mathbf{c}_1 \leqslant \mathbf{u}$ for the elements of $M_{\bar{\mathbf{p}}}$. The existence of such a pair implies that in $\mathcal{R}_\alpha(\leqslant \mathbf{u})$, the degrees below $\mathbf{u}$, $M_{\bar{\mathbf{p}}}$ has no least upper bound. Also, because $M_{\bar{\mathbf{p}}}$ is standard, $M_{\bar{\mathbf{p}}}$ is the only nontrivial initial segment of $M_{\bar{\mathbf{p}}}$ closed under the successor operation.

However, in chapter 4 we show (theorem 4.4) that there is a formula θ such that if α is Σ_2-admissible, $\operatorname{cf}_{\Sigma_3(J_\alpha)}(\alpha) > \omega$ and $\mathbf{u}$ is low$_2$, then for all $\bar{\mathbf{p}}, \bar{\mathbf{e}} \leqslant \mathbf{u}$ which satisfy $\chi_{\texttt{effective}}$ there is some $\mathbf{c} \leqslant \mathbf{u}$ which is the least upper bound for the standard part of $M_{\bar{\mathbf{p}}}$ in the degrees below $\mathbf{u}$; moreover, that standard part is definable as $\theta(\mathcal{R}_\alpha, \mathbf{c}, \bar{\mathbf{p}})$.

We can thus let the formula $\phi_0(y, \mathbf{c}, \bar{\mathbf{p}}, \bar{\mathbf{e}})$ state that $\mathbf{c}, \bar{\mathbf{p}}, \bar{\mathbf{e}} < y$ and that $\chi_{\texttt{effective}}(\bar{\mathbf{p}}, \bar{\mathbf{e}})$ holds, that $\theta(\mathcal{R}_\alpha, \mathbf{c}, \bar{\mathbf{p}})$ is a nontrivial initial segment of $M_{\bar{\mathbf{p}}}$ closed under the successor operation, and that $\mathbf{c}$ is the least upper bound for $\theta(\mathcal{R}_\alpha, \mathbf{c}, \bar{\mathbf{p}})$

in the degrees below y. Let $\phi(y)$ state the existence of some $\bar{\mathbf{p}}, \bar{\mathbf{e}}$ below y such that $\chi_{\texttt{effective}}(\bar{\mathbf{p}}, \bar{\mathbf{e}})$ holds but for no $\mathbf{c}$ below y does $\phi_0(y, \mathbf{c}, \bar{\mathbf{p}}, \bar{\mathbf{e}})$ hold. We thus argued that for a Σ_2-admissible ordinal α such that $\mathrm{cf}_{\Sigma_3(J_\alpha)}(\alpha) = \omega$, $\phi(\mathbf{u})$ holds (in $\mathcal{R}_\alpha$) for all $\mathbf{u}$ which permit promptly; and that if α is Σ_2-admissible such that $\mathrm{cf}_{\Sigma_3(J_\alpha)}(\alpha) > \omega$, then $\neg\phi(\mathbf{u})$ holds for all $\mathbf{u}$ which are low$_2$.

Let X be an additional unary predicate. Let **PS** denote the collection of degrees which permit promptly; let $\mathbf{L}_1$ and $\mathbf{L}_2$ denote the classes of low and low$_2$ degrees respectively. Now there is always a low promptly permitting degree. Together with the results mentioned, we get the following:

THEOREM 1.3. *Let α be a Σ_2-admissible ordinal.*

(1) $(\mathcal{R}_\alpha, \mathbf{PS}) \models \forall y \in X\, \phi(y)$ *iff* $\mathrm{cf}_{\Sigma_3(J_\alpha)}(\alpha) = \omega$.
(2) $(\mathcal{R}_\alpha, \mathbf{L}_1) \models \exists y \in X\, \phi(y)$ *iff* $(\mathcal{R}_\alpha, \mathbf{L}_2) \models \exists y \in X\, \phi(y)$ *iff* $\mathrm{cf}_{\Sigma_3(J_\alpha)}(\alpha) = \omega$.

One would like of course to improve this by eliminating the extra unary predicate; one would think that the most likely candidate is the class of promptly permitting degrees, which is definable in $\mathcal{R}_\omega$. The classical proof can be carried out if, for example, $\varrho^2_\alpha = \omega$, but fails miserably in other cases, and we in fact suspect that in some cases there may be a noncapping degree (i.e. a degree which is not half of a minimal pair) which is not promptly permitting. It follows, though (as no degree which permits promptly can be half of a minimal pair), that if we let X state that y is noncappable then for every Σ_2-admissible ordinal α, if $\mathrm{cf}_{\Sigma_3(J_\alpha)}(\alpha) > \omega$ then $\mathcal{R}_\alpha \models \exists y \in X\, \phi(y)$ and if $\varrho^2_\alpha = \omega$ then $\mathcal{R}_\alpha \models \neg\exists y \in X\, \phi(y)$.

If we are willing to go one level higher to the Σ_3 level, then we get the following (this is also a consequence of theorem 4.4 and the examples given after that theorem).

THEOREM 1.4. *If $\mathrm{cf}_{\Sigma_4(J_\alpha)}(\alpha) > \omega$ and either α is Σ_3-admissible or $\varrho^3_\alpha = \alpha$, then $\mathcal{R}_\alpha \models \neg\exists y\, \phi(y)$.*

This is an elementary difference between such αs and Σ_2-admissible αs such that $\mathrm{cf}_{\Sigma_3(J_\alpha)}(\alpha) = \omega$.

1.3. $\mathcal{R}_\alpha$ is Sometimes Complicated. In chapter 2 we show that for some α, the theory of $\mathcal{R}_\alpha$ is complicated, in terms of its complexity. We show:

THEOREM 1.5. *Let α be an admissible ordinal. If $\varrho^2_\alpha = \omega$ or if α is Σ_2-admissible and $\mathrm{cf}_{\Sigma_3(J_\alpha)}(\alpha) = \omega$ then $\mathcal{O}^{(\omega)} \leqslant_1 \mathrm{Th}(\mathcal{R}_\alpha)$ (where $\mathcal{O}$ is Kleene's complete Π^1_1 set).*

The method of proof is detailed in section 1. The general idea follows the constructions of [**NSS98**]. We show that there is a way to code structures in the language of arithmetic in a uniform way using parameters; and then show that there is a nonempty *correctness condition* on the parameters which is first order and which ensures that the model coded is standard. Further, we show that there is a uniform way to define the isomorphism between any two such models, so that identifying all copies we get a parameterless interpretation of $\mathbb{N}$ in $\mathcal{R}_\alpha$. Further, we show that we can code $\mathcal{O}$ in such a way and get a parameterless definition of $\mathcal{O}$ (in the coded model); this implies the theorem.

Let ψ be the sentence given by theorem 1.1. ψ holds in the classical r.e. degrees. For $\mathcal{R}_\alpha$ we get a dichotomy: if $\varrho^2_\alpha > \omega$ then ψ fails in $\mathcal{R}_\alpha$; and if $\varrho^2_\alpha = \omega$ then $\mathrm{Th}(\mathcal{R}_\alpha)$

is complicated (in particular, it is not hyperarithmetic, whereas $\mathrm{Th}(\mathcal{R}_\omega)$ which has complexity $\mathbf{0}^{(\omega)}$ lies low in the hyperarithmetic hierarchy.) We get the following.

THEOREM 1.6. *For every admissible ordinal α, $\mathcal{R}_\alpha$ and $\mathcal{R}_\omega$ are not elementarily equivalent.*

CHAPTER 2

Coding Into the R.E. Degrees

In this chapter we use the machinery of SW-sets developed in [**NSS98**] and apply it to the α-r.e. degrees for various admissible ordinals α. Using it we code models of arithmetic (with an extra unary predicate) into the α-r.e. degrees, and show how (using the machinery of comparison maps) we can pick out the standard models in which the unary predicate is interpreted as a relatively complicated set (the main example is Kleene's $\mathcal{O}$). For these αs we immediately get that $\mathrm{Th}(\mathcal{R}_\alpha)$ is at least as complicated as the set coded (and in particular is not hyperarithmetic and not equal to $\mathrm{Th}(\mathcal{R}_\omega)$).

1. The Coding

Fix an admissible ordinal α and work in the α-r.e. degrees, $\mathcal{R}_\alpha$. We repeat the coding machinery which is developed in [**NSS98**].

DEFINITION. The *SW set* defined by a quadruple of parameters $\bar{\mathbf{p}} = (\mathbf{p}, \mathbf{q}, \mathbf{r}, \mathbf{l})$ (denoted by $G_{\bar{\mathbf{p}}}$) is the collection of $\mathbf{g} < \mathbf{r}$ which are minimal solutions of the inequality $\mathbf{g} \vee \mathbf{p} \geqslant \mathbf{q}$. For $\mathbf{g}_0, \mathbf{g}_1 \in G_{\bar{\mathbf{p}}}$, we let $\mathbf{g}_0 \leqslant_{\bar{\mathbf{p}}} \mathbf{g}_1$ if $\mathbf{g}_0 \leqslant \mathbf{g}_1 \vee \mathbf{l}$.

REMARK 2.1. $G_{\bar{\mathbf{p}}}$ may be empty, finite or infinite, but if not empty is always an antichain of degrees. $\leqslant_{\bar{\mathbf{p}}}$ is always a pre-partial ordering on $G_{\bar{\mathbf{p}}}$.

Next, we decode structures for the language of arithmetic from the partial ordering $\leqslant_{\bar{\mathbf{p}}}$. Let $\bar{\mathbf{p}}$ be a quadruple of parameters. In what follows we define several relations on $G_{\bar{\mathbf{p}}}$ and describe conditions on the sets and relations defined. All of these conditions together are taken as a *correctness condition* $\chi_{\mathtt{SW}}(\bar{\mathbf{p}})$ on the parameters (and indeed we note that all conditions are expressible in a first order way in the language of partial orderings, so that $\chi_{\mathtt{SW}}(\mathcal{R}_\alpha)$ is definable in $(\mathcal{R}_\alpha, \leqslant_\alpha)$.)

(1) $G_{\bar{\mathbf{p}}}$ is nonempty and $\leqslant_{\bar{\mathbf{p}}}$ is a partial ordering on $G_{\bar{\mathbf{p}}}$.
(2) $M_{\bar{\mathbf{p}}}$ is defined to be the collection of $\leqslant_{\bar{\mathbf{p}}}$-minimal elements. $M_{\bar{\mathbf{p}}}$ is nonempty.
(3) $\mathtt{pair}_{\bar{\mathbf{p}}}(\mathbf{a}, \mathbf{b}, \mathbf{c})$ holds if $\mathbf{a}, \mathbf{b} \in M_{\bar{\mathbf{p}}}$ and there is a 2-chain (in $\leqslant_{\bar{\mathbf{p}}}$) between $\mathbf{a}$ and $\mathbf{c}$ and a 3-chain between $\mathbf{b}$ and $\mathbf{c}$. $\mathtt{pair}_{\bar{\mathbf{p}}}$ defines a total function from $M^2_{\bar{\mathbf{p}}}$ to $G_{\bar{\mathbf{p}}}$ and so we write $\mathbf{c} = \mathtt{pair}_{\bar{\mathbf{p}}}(\mathbf{a}, \mathbf{b})$.
(4) $+_{\bar{\mathbf{p}}}(\mathbf{a}, \mathbf{b}, \mathbf{c})$ holds if $\mathbf{a}, \mathbf{b}, \mathbf{c} \in M_{\bar{\mathbf{p}}}$ and there is a 4-chain in $\leqslant_{\bar{\mathbf{p}}}$ between $\mathbf{c}$ and $\mathtt{pair}_{\bar{\mathbf{p}}}(\mathbf{a}, \mathbf{b})$. $+_{\bar{\mathbf{p}}}$ defines a total function on $M^2_{\bar{\mathbf{p}}}$; we write $\mathbf{c} = \mathbf{a} +_{\bar{\mathbf{p}}} \mathbf{b}$.
(5) $\times_{\bar{\mathbf{p}}}(\mathbf{a}, \mathbf{b}, \mathbf{c})$ holds if $\mathbf{a}, \mathbf{b}, \mathbf{c} \in M_{\bar{\mathbf{p}}}$ and there is a 5-chain in $\leqslant_{\bar{\mathbf{p}}}$ between $\mathbf{c}$ and $\mathtt{pair}_{\bar{\mathbf{p}}}(\mathbf{a}, \mathbf{b})$. $\times_{\bar{\mathbf{p}}}$ defines a total function on $M^2_{\bar{\mathbf{p}}}$; we write $\mathbf{c} = \mathbf{a} \times_{\bar{\mathbf{p}}} \mathbf{b}$.
(6) $(M_{\bar{\mathbf{p}}}, +_{\bar{\mathbf{p}}}, \times_{\bar{\mathbf{p}}})$ satisfies some sufficiently rich finite fragment of arithmetic (say Robinson arithmetic); the important fact is that $M_{\bar{\mathbf{p}}}$ always has a standard initial segment. As is a usual practice, if φ is a formula in first

order arithmetic then we let $\varphi(M_{\bar{\mathbf{p}}})$ or $\varphi^{M_{\bar{\mathbf{p}}}}$ denote its interpretation in $(M_{\bar{\mathbf{p}}}, +_{\bar{\mathbf{p}}}, \times_{\bar{\mathbf{p}}})$. We thus get $n^{M_{\bar{\mathbf{p}}}}$ for each $n < \omega$, $\leqslant^{M_{\bar{\mathbf{p}}}}$ (not to be confused with $\leqslant_{\bar{\mathbf{p}}}$), etc.

(7) We let $X_{\bar{\mathbf{p}}}$ be the collection of $\mathbf{a} \in M_{\bar{\mathbf{p}}}$ such that there is a 6-chain in $\leqslant_{\bar{\mathbf{p}}}$ above $\mathbf{a}$.

REMARK 2.2. Let $X \subset \omega$. Then we can easily produce a partial ordering $\preccurlyeq_X$ which codes the standard model of arithmetic with extra predicate X. That is, there is some partial ordering $\preccurlyeq_X$ on ω (which is effectively computed from X) such that for all $\bar{\mathbf{p}}$, if $\chi_{\mathtt{SW}}(\bar{\mathbf{p}})$ holds and $(G_{\bar{\mathbf{p}}}, \leqslant_{\bar{\mathbf{p}}}) \cong (\omega, \preccurlyeq_X)$ then $(M_{\bar{\mathbf{p}}}, +_{\bar{\mathbf{p}}}, \times_{\bar{\mathbf{p}}})$ is isomorphic to the standard model of arithmetic and $X = \{n < \omega : n^{M_{\bar{\mathbf{p}}}} \in X_{\bar{\mathbf{p}}}\}$.

We proceed along with [**NSS98**] to construct comparison maps. We first spell out the necessary recursion theoretic facts that are used.

For the rest of the section, let α be an admissible ordinal which satisfies one of the following:

- $\varrho^2_\alpha = \omega$.
- α is Σ_2-admissible and $\mathrm{cf}_{\Sigma_3(J_\alpha)}(\alpha) = \omega$.

See appendix A for the definitions of Σ_2-admissibility and the Σ_n-cofinality, and appendix C for the definition of the Σ_2-projectum ϱ^2_α.

For such α, we will prove the following theorems:

THEOREM 2.3. *Let $\preccurlyeq$ be an α-recursive partial ordering on ω, and let $\mathbf{a} > \mathbf{0}$ be a nonzero α-r.e degree. Then there is a tuple $\bar{\mathbf{p}} = (\mathbf{p}, \mathbf{q}, \mathbf{r}, \mathbf{l})$ such that $\mathbf{r}$ is low, $\mathbf{r} \leqslant \mathbf{a}$, and such that $(\omega, \preccurlyeq) \cong (G_{\bar{\mathbf{p}}}, \leqslant_{\bar{\mathbf{p}}})$.*

If $\mathbf{r}$ is low we also say that the SW set $G_{\bar{\mathbf{p}}}$ is low. If $\mathbf{r} \leqslant \mathbf{a}$ then we say that the SW set $G_{\bar{\mathbf{p}}}$ is *coded below* $\mathbf{a}$. If $(\omega, \preccurlyeq) \cong (G_{\bar{\mathbf{p}}}, \leqslant_{\bar{\mathbf{p}}})$ we say that $G_{\bar{\mathbf{p}}}$ codes $\preccurlyeq$.

THEOREM 2.4. *Let $\preccurlyeq$ be an α-recursive partial ordering on ω and let H be an α-recursive set of $\preccurlyeq$-minimal elements. Also let $\langle \mathbf{u}_i \rangle_{i \in H}$ be a sequence of uniformly α-r.e. degrees, and $\langle \mathbf{v}_{i,j} \rangle_{i \in H, j < \omega}$ be an array of uniformly α-r.e., uniformly low degrees such that for all $i \in H$ and $j < \omega$, $\mathbf{u}_i \nleqslant \mathbf{v}_{i,j}$. Then there is a tuple $\bar{\mathbf{p}} = (\mathbf{p}, \mathbf{q}, \mathbf{r}, \mathbf{l})$ such that $\mathbf{r}$ is low and such that we can enumerate $G_{\bar{\mathbf{p}}} = \{\mathbf{g}_i : i < \omega\}$, such that $(\omega, \preccurlyeq) \cong (G_{\bar{\mathbf{p}}}, \leqslant_{\bar{\mathbf{p}}})$ by the isomorphism $i \longrightarrow \mathbf{g}_i$ and such that for $i \in H$ and $j < \omega$, $\mathbf{g}_i \leqslant \mathbf{u}_i$ and $\mathbf{g}_i \nleqslant \mathbf{v}_{i,j}$.*

In the above theorems and in the following arguments we use the notion of lowness, which is discussed in appendix B.

We first show that there is a uniform way to define *comparison maps* from models $M_{\bar{\mathbf{p}}_0} \to M_{\bar{\mathbf{p}}_1}$ if both models are low and $M_{\bar{\mathbf{p}}_0}$ is standard.

DEFINITION. Let $\bar{\mathbf{p}}_0, \bar{\mathbf{p}}_1$ be tuples such that $\chi_{\mathtt{SW}}(\bar{\mathbf{p}}_i)$ holds for $i < 2$. A function h is a *comparison map* from $M_{\bar{\mathbf{p}}_0}$ to $M_{\bar{\mathbf{p}}_1}$ if it is a 1-1 function from an initial segment of $M_{\bar{\mathbf{p}}_0}$ onto an initial segment of $M_{\bar{\mathbf{p}}_1}$ preserving the least element (i.e. $h(0^{M_{\bar{\mathbf{p}}_0}}) = 0^{M_{\bar{\mathbf{p}}_1}}$) and the successor relation.

LEMMA 2.5. *There is a formula $\mathtt{map}_{\mathtt{low}}(x, y, \bar{w}, \bar{z})$ such that for all tuples $\bar{\mathbf{p}}_0, \bar{\mathbf{p}}_1$ satisfying $\chi_{\mathtt{SW}}$ such that $\mathbf{r}_0, \mathbf{r}_1$ are low and $M_{\bar{\mathbf{p}}_0}$ is isomorphic to the standard model of arithmetic, $\mathtt{map}_{\mathtt{low}}(x, y, \bar{\mathbf{p}}_0, \bar{\mathbf{p}}_1)$ defines a comparison map from $M_{\bar{\mathbf{p}}_0}$ to $M_{\bar{\mathbf{p}}_1}$ which is total on $M_{\bar{\mathbf{p}}_0}$.*

If $\chi_{\mathtt{SW}}(\bar{\mathbf{p}})$ holds and $M_{\bar{\mathbf{p}}}$ is isomorphic to the standard model of arithmetic then we say that $M_{\bar{\mathbf{p}}}$ *is standard*. If $\chi_{\mathtt{SW}}(\bar{\mathbf{p}})$ holds and $G_{\bar{\mathbf{p}}}$ is low then we also say that $M_{\bar{\mathbf{p}}}$ is low; if $G_{\bar{\mathbf{p}}}$ is coded below $\mathbf{a}$ then we also say that $M_{\bar{\mathbf{p}}}$ is coded below $\mathbf{a}$.

PROOF. The formula $\mathtt{map}_{\mathtt{low}}$ is defined as follows. Let $\psi(x, y, \bar{u}, \bar{v}, \bar{z}, t)$ state that $\chi_{\mathtt{SW}}$ holds for $\bar{u}, \bar{v}$ and $\bar{z}$; that $x \in M_{\bar{u}}$, $y \in M_{\bar{v}}$ and $t \in M_{\bar{z}}$; and that there is some $s \in M_{\bar{z}}$ such that $s <^{M_{\bar{z}}} t$ and such that

(1) x is the least element of $M_{\bar{u}}$ (according to $<^{M_{\bar{u}}}$) such that $s \leqslant x$; and
(2) y is the least element of $M_{\bar{v}}$ (according to $<^{M_{\bar{v}}}$) such that $(s +_{\bar{z}} t) \leqslant x$.

We let $\mathtt{map}_{\mathtt{low}}(x, y, \bar{u}, \bar{w})$ state the existence of some $\bar{z}$ and t such that

$$\psi(-, -, \bar{u}, \bar{v}, \bar{z}, t)$$

defines a comparison map from $M_{\bar{u}}$ to $M_{\bar{v}}$, and such that this map takes x to y.

Now it is clear that if $M_{\bar{\mathbf{p}}_0}$ is standard then for any $\bar{\mathbf{p}}_1$ (satisfying $\chi_{\mathtt{SW}}$ of course), $\mathtt{map}_{\mathtt{low}}(x, y, \bar{\mathbf{p}}_0, \bar{\mathbf{p}}_1)$ defines a comparison map from $M_{\bar{\mathbf{p}}_0}$ to $M_{\bar{\mathbf{p}}_1}$. We wish to show that if in addition both $M_{\bar{\mathbf{p}}_0}$ and $M_{\bar{\mathbf{p}}_1}$ are low then $\mathtt{map}_{\mathtt{low}}(x, y, \bar{\mathbf{p}}_0, \bar{\mathbf{p}}_1)$ is total. Fix such $\bar{\mathbf{p}}_0$ and $\bar{\mathbf{p}}_1$. It suffices to show, for every $n < \omega$, the existence of some $\bar{\mathbf{p}}$ such that $\chi_{\mathtt{SW}}(\bar{\mathbf{p}})$ holds and such that the further following property holds:

- For all $i, j < n$, $i^{M_{\bar{\mathbf{p}}}} \leqslant j^{M_{\bar{\mathbf{p}}_0}}$ iff $i = j$ and $(i+n)^{M_{\bar{\mathbf{p}}}} \leqslant j^{M_{\bar{\mathbf{p}}_1}}$ iff $i = j$.

The existence of such $\bar{\mathbf{p}}$ (we can even require that $M_{\bar{\mathbf{p}}}$ be standard) follows from theorem 2.4 by making the following substitutions. We let $\preccurlyeq$ be the usual partial ordering which codes the standard model of arithmetic (see remark 2.2). For $m < \omega$ let h_m be the element of $\preccurlyeq$ coding m and let $H = \{h_m : m < 2n\}$. For $m < n$ we let $\mathbf{u}_{h_m} = m^{M_{\bar{\mathbf{p}}_0}}$ and $\mathbf{u}_{h_{m+n}} = m^{M_{\bar{\mathbf{p}}_1}}$. For $i, j < n$ such that $i \neq j$, let $\mathbf{v}_{h_m, j} = j^{M_{\bar{\mathbf{p}}_0}}$ and $\mathbf{v}_{h_{m+n}, j} = j^{M_{\bar{\mathbf{p}}_0}}$; for all other $j < \omega$, let $\mathbf{v}_{h_i, j} = \mathbf{v}_{h_{i+n}, j} = \mathbf{0}$. The condition $\mathbf{u}_i \nleqslant \mathbf{v}_{i,j}$ is fulfilled because $M_{\bar{\mathbf{p}}_0}$ and $M_{\bar{\mathbf{p}}_1}$ are antichains of α-r.e. degrees. Finally, the (uniform) lowness condition follows from the lowness of $M_{\bar{\mathbf{p}}_0}$ and $M_{\bar{\mathbf{p}}_1}$.

Fix $n < \omega$ and get $\bar{\mathbf{p}}$ as described.

It is clear that for all $m < n$, $\psi(m^{M_{\bar{\mathbf{p}}_0}}, m^{M_{\bar{\mathbf{p}}_1}}, \bar{\mathbf{p}}_0, \bar{\mathbf{p}}_1, \bar{\mathbf{p}}, n^{M_{\bar{\mathbf{p}}}})$ holds. What we need is to show that $\psi(x, y, \bar{\mathbf{p}}_0, \bar{\mathbf{p}}_1, \bar{\mathbf{p}}, n^{M_{\bar{\mathbf{p}}}})$ defines a comparison map from $M_{\bar{\mathbf{p}}_0}$ to $M_{\bar{\mathbf{p}}_1}$. But in fact it is not difficult to see that for all x and y, $\psi(x, y, \bar{\mathbf{p}}_0, \bar{\mathbf{p}}_1, \bar{\mathbf{p}}, n^{M_{\bar{\mathbf{p}}}})$ holds iff for some $m < n$ we have $x = m^{M_{\bar{\mathbf{p}}_0}}$ and $y = m^{M_{\bar{\mathbf{p}}_1}}$. For suppose that $\psi(x, y, \bar{\mathbf{p}}_0, \bar{\mathbf{p}}_1, \bar{\mathbf{p}}, n^{M_{\bar{\mathbf{p}}}})$ holds; let $s \in M_{\bar{\mathbf{p}}}$ witnesses this fact. ψ implies that $s <^{M_{\bar{\mathbf{p}}}} n^{M_{\bar{\mathbf{p}}}}$ and so $s = m^{M_{\bar{\mathbf{p}}}}$ for some $m < n$. Then $s \leqslant m^{M_{\bar{\mathbf{p}}_0}}, x$ and both are minimal with respect to that property, so $x = m^{M_{\bar{\mathbf{p}}_0}}$; similarly, $y = m^{M_{\bar{\mathbf{p}}_1}}$. $\square$

We now get a correctness condition strengthening $\chi_{\mathtt{SW}}$ which implies that the model coded is standard, and we get comparison maps between the models $M_{\bar{\mathbf{p}}}$ satisfying this condition.

THEOREM 2.6. *There is a formula $\chi_{\mathtt{standard}}(\bar{\mathbf{p}})$ (implying $\chi_{\mathtt{SW}}(\bar{\mathbf{p}})$) such that for all $\bar{\mathbf{p}}$ satisfying $\chi_{\mathtt{standard}}$, $M_{\bar{\mathbf{p}}}$ is standard. There is a formula $\mathtt{map}$ such that for every $\bar{\mathbf{p}}_0, \bar{\mathbf{p}}_1$ satisfying $\chi_{\mathtt{standard}}$, $\mathtt{map}(x, y, \bar{\mathbf{p}}_0, \bar{\mathbf{p}}_1)$ defines the isomorphism between $M_{\bar{\mathbf{p}}_0}$ and $M_{\bar{\mathbf{p}}_1}$. Further, if $X \subset \omega$ is α-recursive, then there is some $\bar{\mathbf{p}}$ such that $\chi_{\mathtt{standard}}(\bar{\mathbf{p}})$ holds and $M_{\bar{\mathbf{p}}}$ codes X.*

PROOF. $\chi_{\texttt{standard}}(\bar{\mathbf{p}})$ states that $\chi_{\texttt{SW}}(\bar{\mathbf{p}})$ holds and that for every other tuple $\bar{\mathbf{p}}' = (\mathbf{p}', \mathbf{q}', \mathbf{r}', \mathbf{l}')$ satisfying $\chi_{\texttt{SW}}$ such that $\mathbf{r}' \leqslant \mathbf{r}$, $\texttt{map}_{\texttt{low}}(x, y, \bar{\mathbf{p}}, \bar{\mathbf{p}}')$ defines a total map on $M_{\bar{\mathbf{p}}}$.

Suppose that $\chi_{\texttt{standard}}(\bar{\mathbf{p}})$ holds. By theorem 2.3, we know that there is some $\bar{\mathbf{p}}'$ such that $\mathbf{r}' \leqslant \mathbf{r}$ and such that $M_{\bar{\mathbf{p}}'}$ is standard. Totality of the comparison map from $M_{\bar{\mathbf{p}}}$ into $M_{\bar{\mathbf{p}}'}$ implies that $M_{\bar{\mathbf{p}}}$ is standard as well.

To define $\texttt{map}$, we note that in the proof of lemma 2.5, we say that whenever $M_{\bar{\mathbf{p}}}$ is standard and $M_{\bar{\mathbf{p}}'}$ is any other coded model, $\texttt{map}_{\texttt{low}}(x, y, \bar{\mathbf{p}}, \bar{\mathbf{p}}')$ defines a comparison map from $M_{\bar{\mathbf{p}}}$ to $M_{\bar{\mathbf{p}}'}$, not necessarily total. For such $\bar{\mathbf{p}}$ and $\bar{\mathbf{p}}'$, let $h_{\bar{\mathbf{p}},\bar{\mathbf{p}}'}$ denote this comparison map.

Now let $\bar{\mathbf{p}}_0, \bar{\mathbf{p}}_3$ be two tuples satisfying $\chi_{\texttt{standard}}$. We let $\texttt{map}(x, y, \bar{\mathbf{p}}_0, \bar{\mathbf{p}}_3)$ state that there are $\bar{\mathbf{p}}_1, \bar{\mathbf{p}}_2$, also satisfying $\chi_{\texttt{standard}}$, such that

$$y = h^{-1}_{\bar{\mathbf{p}}_3,\bar{\mathbf{p}}_2} \circ h_{\bar{\mathbf{p}}_1,\bar{\mathbf{p}}_2} \circ h_{\bar{\mathbf{p}}_0,\bar{\mathbf{p}}_1}(x).$$

An inverse of a comparison map, a composition of comparison maps, and the union of comparison maps, is a comparison map, if the domain is standard. Thus $\texttt{map}(x, y, \bar{\mathbf{p}}_0, \bar{\mathbf{p}}_3)$ defines a comparison map from $M_{\bar{\mathbf{p}}_0}$ to $M_{\bar{\mathbf{p}}_3}$; we need to see that this is always total. Well, theorem 2.3 implies the existence of standard low $M_{\bar{\mathbf{p}}_1}$ and $M_{\bar{\mathbf{p}}_2}$ such that $\mathbf{r}_1 \leqslant \mathbf{r}_0$ and $\mathbf{r}_2 \leqslant \mathbf{r}_3$. Lowness of $M_{\bar{\mathbf{p}}_1}$ and $M_{\bar{\mathbf{p}}_2}$ implies that $\chi_{\texttt{standard}}(\bar{\mathbf{p}}_i)$ holds for $i = 1, 2$; this follows from lemma 2.5.

Now totality of $h_{\bar{\mathbf{p}}_0,\bar{\mathbf{p}}_1}$ and $h_{\bar{\mathbf{p}}_3,\bar{\mathbf{p}}_2}$ follows from the fact that $\chi_{\texttt{standard}}(\bar{\mathbf{p}}_i)$ holds for $i = 0, 3$; totality of $h_{\bar{\mathbf{p}}_1,\bar{\mathbf{p}}_2}$ follows from the fact that $M_{\bar{\mathbf{p}}_1}$ is low and standard.

The last assertion of the theorem follows immediately from theorem 2.3; again any low, standard model satisfies $\chi_{\texttt{standard}}$. □

Theorem 2.6 enables us to interpret in $\mathcal{R}_\alpha$, without parameters, an ω-model $M = (\mathbb{N}_M, \mathbb{R}_M)$ of *second* order arithmetic. The first order part $\mathbb{N}_M$ is given by identifying all copies $M_{\bar{\mathbf{p}}}$ of the standard model of arithmetic for $\bar{\mathbf{p}}$ satisfying $\chi_{\texttt{standard}}$, under the isomorphisms given by the formula $\texttt{map}$. In detail, let $\bar{\mathbf{p}}, \bar{\mathbf{p}}'$ satisfy $\chi_{\texttt{standard}}$, and let $x \in M_{\bar{\mathbf{p}}}$, $y \in M_{\bar{\mathbf{p}}'}$. Let $(x, \bar{\mathbf{p}}) \sim_{\mathbb{N}} (y, \bar{\mathbf{p}}')$ if $\texttt{map}(x, y, \bar{\mathbf{p}}, \bar{\mathbf{p}}')$ holds. $\mathbb{N}_M$ is the collection of $\sim_{\mathbb{N}}$-equivalence classes. As $\texttt{map}$ defines isomorphisms, $\sim_{\mathbb{N}}$ is a *congruence* relation for the operations of arithmetic, and so the arithmetical structure of $\mathbb{N}_M$ is also definable in $\mathcal{R}_\alpha$.

$\mathbb{R}_M$ consists of all sets which are coded as $X_{\bar{\mathbf{p}}}$ for $\bar{\mathbf{p}}$ satisfying $\chi_{\texttt{standard}}$. Namely, for $\bar{\mathbf{p}}, \bar{\mathbf{p}}'$ satisfying $\chi_{\texttt{standard}}$, let $\bar{\mathbf{p}} \sim_{\mathbb{R}} \bar{\mathbf{p}}'$ if for all $n < \omega$, $n^{M_{\bar{\mathbf{p}}}} \in X_{\bar{\mathbf{p}}}$ iff $n^{M_{\bar{\mathbf{p}}'}} \in X_{\bar{\mathbf{p}}'}$. This relation is definable in $\mathcal{R}_\alpha$. We let $\mathbb{R}_M$ be the set of $\sim_{\mathbb{R}}$-equivalence classes.

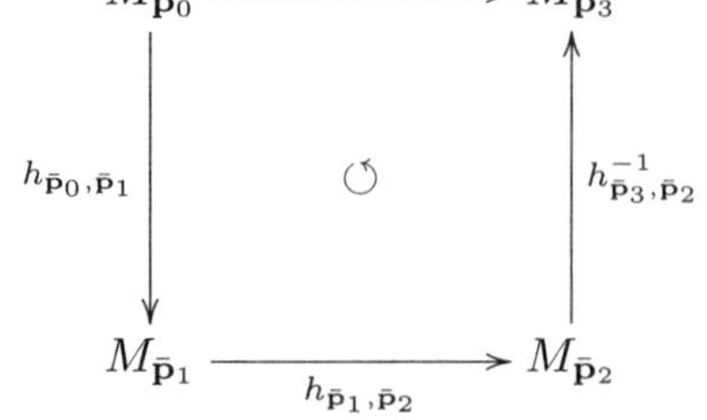

FIGURE 2.1. The definition of $\texttt{map}$

For $n < \omega$ and $Y \in \mathbb{R}_M$, $n \in^M Y$ if for some (any) $\bar{\mathbf{p}} \in Y$, $n^{M_{\bar{\mathbf{p}}}} \in X_{\bar{\mathbf{p}}}$. This too is definable in $\mathcal{R}_\alpha$. We can quantify over $\mathbb{R}_M$ by quantifying over such $\bar{\mathbf{p}}$.

PROOF OF THEOREM 1.5. Let $A \subset \omega$ be any Σ^1_1 set and suppose that $\alpha > \omega_1^{CK}$. In this case, A is α-finite, and so there is some $M_{\bar{\mathbf{p}}}$ which codes A; so $A \in \mathbb{R}_M$. Let ψ be an arithmetical formula such that for all $n < \omega$, $n \in A$ iff there is some $X \subset \omega$ such that $\psi(n, X)$. We actually know that if $n \in A$ then there is some $X \leqslant_T \mathcal{O}$ such that $\psi(n, X)$ holds (this is because the witness X can be taken to be any infinite path in an ω-splitting recursive tree, so we can take the leftmost one). We have $X \in \mathbb{R}_M$, and ψ is absolute for ω-models of second order arithmetic, and so A is definable in M. It follows that $\mathrm{Th}(\mathbb{N}, A) \leqslant_1 \mathrm{Th}(\mathcal{R}_\alpha)$. If A is Σ^1_1-complete then $\mathcal{O}^{(\omega)} \equiv_1 \mathrm{Th}(\mathbb{N}, A)$. □

A VERY SIMILAR PROOF OF THEOREM 1.5, FOLLOWING [**GSS**]. By the same argument, $\mathcal{O} \in \mathbb{R}_M$. Further, there is some arithmetical formula ψ such that $\mathcal{O}$ is inclusion-wise the least solution of ψ (see [**Sac90**, p.8]). Again ψ is absolute for M and set inclusion is also definable in M and so this definition of $\mathcal{O}$ holds in M. □

The case $\alpha = \omega_1^{CK}$ has to be treated slightly differently (as is done in [**GSS**]) because $\mathcal{O}$ is not ω_1^{CK}-recursive. Nevertheless even for $\alpha = \omega_1^{CK}$ we get $\mathcal{O} \in \mathbb{R}_M$ (In fact $\Delta_2(J_\alpha) \cap \mathbb{R} \subset \mathbb{R}_M$) and the previous arguments hold. We give more details about the required recursion-theoretic constructions (i.e. the proofs of theorems 2.3 and 2.4 where "$\preccurlyeq$ is α-recursive" replaced by "$\preccurlyeq$ is $\Delta_2(J_\alpha)$") in section 3.1.

In fact, if $\alpha > \omega_1^{CK}$ then iterating this coding shows we can recover more complicated sets. If α is the nth admissible ordinal then these techniques show that $\mathcal{O}^{\langle n \rangle}$, the nth iterate of the hyperjump, is in $\mathbb{R}_M$. As $\mathcal{O}^{\langle n+1 \rangle}$ is definable from $\mathcal{O}^{\langle n \rangle}$ in much the same way that $\mathcal{O}$ was, we can recover it definably in the model interpreted in $\mathcal{R}_\alpha$. This yields elementary differences between the structures of α-r.e. degrees for these ordinals.

This process of coding and recovery can be carried into the transfinite; for example, if α is the ωth admissible ordinal then $\mathcal{O}^{\langle \omega \rangle}$ is α-finite and so can be coded; it can be recovered by the inductive formula which recovers each of its columns.

2. A Template for the Constructions

The proofs of theorems 2.3 and 2.4 are done by priority constructions which depend on the properties of the underlying admissible ordinal α. To avoid repetition and to add rigor, we first present a general template to fit all of these constructions, and prove general facts which are shared by all constructions. Familiarity with the techniques of constructing SW sets in the ω-r.e. degrees, as presented in [**NSS98**], would be helpful, but we give a complete proof.

From now on we omit the prefix α from the terms 'recursive' and 'r.e.'. Again we remind the reader that the definitions of various notions used in admissible recursion theory (such as strong, weak or nice functionals) are given in appendix A.

2.1. The Requirements. We are given a recursive partial ordering $\preccurlyeq$ on ω. We construct sets G_n, P, Q and L. We let $R = \oplus G_n$. We strive to meet the following requirements.

(1) T_n: $Q \leqslant_\alpha G_n \oplus P$.

(2) $M_{\Theta,\Phi,W}$: If $\Theta(R) = W$ and $\Phi(W \oplus P) = Q$ then there is some j such that $G_j \leqslant_\alpha W$. Here (Θ, Φ, W) ranges over triples consisting of a strong functional, a weak functional and an r.e. set.
(3) $D_{i,j,\Psi}$: If $j \neq i$, then $\Psi(G_i) \neq G_j$. Here Ψ ranges over weak functionals.
(4) $N_{i,j,\Psi}$: If $j \not\preccurlyeq i$ then $\Psi(G_i \oplus L) \neq G_j$. Ψ ranges over weak functionals.
(5) $K_{\Xi,K}$: If for all $x \in K$ there are unboundedly many s such that $x \in \Xi(R \oplus P \oplus Q \oplus L)\,[s]$ then from some s onward, $K \subset \Xi(R \oplus P \oplus Q \oplus L)\,[s]$ is correct. Here Ξ ranges over enumeration functionals, including ones appearing in the construction with approximations as built in the construction. K ranges over all α-finite sets which have a greatest element.

We further ensure that if $j \preccurlyeq i$ then $G_j \leqslant_\alpha G_i \oplus L$.

Constructions proving theorem 2.3 ('type 1 constructions') are also given a nonrecursive r.e. set A (which by Sacks's theorem we may assume is amenable), and try to ensure that $R \leqslant_\alpha A$.

Constructions proving theorem 2.4 ('type 2 constructions') are also given a recursive set of $\preccurlyeq$-minimal elements H, and uniformly r.e., uniformly amenable arrays $\langle U_i \rangle_{i \in H}$ and $\langle V_{i,j} \rangle_{i \in H, j < \omega}$, the latter being also uniformly low, such that for all $i \in H$ and $j < \omega$, $U_i \not\leqslant_\alpha V_{i,j}$. We add the requirement

(6) $Z_{i,j,\Psi}$: If $i \in H$, then $G_i \neq \Psi(V_{i,j})$. Ψ ranges over weak functionals.

We may assume that the enumeration given for $V_{i,j}$ is low for Ψ (see appendix B).

We also assume that all functionals given to us are nice (again, see appendix A).

2.2. Elements of the Constructions. All constructions will use a tree of strategies. Each node on the tree (we can always think of the node as an α-finite sequence of ordinals below α) is an *agent*, which at various stages of the construction may be assigned requirements for which they work. The lexicographic ordering on the tree is the priority ordering (so an agent η is stronger than agent ρ if $\eta \subsetneq \rho$ or if η lies on the tree to the left of ρ (lexicographically).) The path of agents *accessible* at stage s of the construction is always defined, is well-ordered by inclusion and is denoted by $\delta\,[s]$.

2.2.1. *Initialization.* The way agents impose finitary-type restraint is by initializing all weaker agents. Whenever an agent enumerates a number into a set or whenever it declares victory, it initializes all weaker agents. Also, at the end of stage s we initialize all agents which lie to the right of $\delta\,[s]$. Whenever an agent is initialized, all weaker agents are initialized as well.

We also ensure that a change in the requirement to which an agent is assigned entails the initialization of the agent: if $\eta \subset \delta\,[s]$ and there is no $t < s$ such that $\eta \subset \delta\,[t]$ and η is assigned to a fixed requirement from t to s, then we initialize η at s.

When an agent gets initialized, we cancel all of its followers, chits and pointers (these are various numbers whose proper definitions will shortly follow). All functionals defined by the agent are abandoned. Of course, if a stage s is a limit of stages at which an agent was initialized, then the agent starts afresh at s as well.

We let

$$\mathtt{init}(\eta)[s] = \sup\{t < s \,:\, \eta \text{ is initialized at } t\}.$$

2.2.2. *Agreement, Expansion and Restraint.* Let $M = M_{\Theta,\Phi,W}$ be a minimality requirement. A number y is *M-confirmed* at s if

$$\Phi(W \oplus P, y) \downarrow = Q(y)[s]$$

with use $\sigma \oplus \pi = \phi(y)[s]$, and

$$\sigma \subset \Theta(R)[s].$$

The length of agreement is

$$\ell(M)[s] = \max\{z \mid \forall y < z, y \text{ is } M\text{-confirmed at } s\}.$$

Suppose that at s, agent $\eta \in \delta[s]$ is working for M. A stage s is *η-expansionary* if for all $t < s$ such that $\eta \in \delta[t]$, $\ell(M)[t] < \ell(M)[s]$.

At stage s, if agent η is working for M, we let

$$r(\eta)[s] = \begin{cases} 0 & \text{if } s \text{ is } \eta\text{-expansionary or a limit of } \eta\text{-expansionary stages,} \\ t & \text{if not, and } t = \sup\{u < s : u \text{ is } \eta\text{-expansionary}\}. \end{cases}$$

We let

$$\texttt{Rest}(\eta)[s] = \sup_{\rho \subsetneq \eta} r(\rho)[s]$$

(agents ρ which lie to the left of η impose restraint by was of initialization so η does not have to take $r(\rho)$ into consideration for such ρ).

2.2.3. *T Requirements.* Suppose that agent η is assigned to T_n. The agent η defines a functional Γ_η, with the intention of having $\Gamma_\eta(G_n \oplus P) = Q$. η may only add an axiom to Γ_η when it is accessible.

All uses of Γ functionals are successor ordinals; $\gamma_\eta(x) = \gamma_\eta(G_n \oplus P, x)$ denotes the length of the use of the computation $\Gamma_\eta(G_n \oplus P, x)$ minus one, so enumerating it into G_n or P destroys the computation.

The following is important.

DEFINITION. When we design the tree and assign requirements to agents, we always make sure that for every agent η (and at every stage s), only finitely many agents $\rho \subset \eta$ are assigned to some T_n requirement at s. We let $n(\rho)[s] < \omega$ be the greatest n such that some $\eta \subset \rho$ is assigned to T_n (at s).

2.2.4. *Followers.* Agents working for D, N or Z requirements may have *followers*. A follower $x < \alpha$ for an agent η working for $D_{i,j,\Psi}$ is *targeted* for G_j and for L. The follower x is *realized* at s if

$$\Psi(G_i, x) \downarrow = 0\,[s].$$

If η works for $N_{i,j,\Psi}$ then x is targeted for G_k for every k such that $j \preccurlyeq k$ (including of course j itself). In this case x is realized if

$$\Psi(G_i \oplus L, x) \downarrow = 0\,[s].$$

In type 2 constructions, a follower x for an agent η working for $Z_{i,j,\Psi}$ is targeted for G_i and for L; it is realized if

$$\Psi(V_{i,j}, x) \downarrow = 0\,[s].$$

In type 1 constructions, a follower x (for any agent) is *permitted* at s if some number $y < x$ enters A at s.

In type 2 constructions, if x is a follower targeted for G_i and $i \in H$ then x is permitted at s if some number $y < x$ enters U_i at s. In these constructions, if x is not targeted for G_i for any $i \in H$ then x is always permitted. Note that by our instructions, and by the fact that H is a set of $\preccurlyeq$-minimal elements, a follower x may be targeted for G_i for at most one $i \in H$.

2.2.5. *Pointers.* An agent η working for a requirement $K_{\Xi,K}$ keeps a *pointer* $i(\eta)[s]$, which is the next element of K of which it needs to take care. Unless initialized, at a limit stage s we have

$$i(\eta)[s] = \min(K \setminus \sup\{i(\eta)[t]\}_{t<s}).$$

When η is initialized we set $i(\eta) = \min K$. If η acts at s then we update the pointer:

$$i(\eta)[s+1] = \min(K \setminus i(\eta)[s]+1).$$

If $i(\eta)[s] = \max K$ and η acts at s then η declares victory.

2.2.6. *Chits.* Suppose that agent η works for $M = M_{\Theta,\Phi,W}$. η defines weak functionals $\Delta_{\eta,j}$ for every $j \leqslant n(\eta)$, with intended oracle W; the intention is that if the hypothesis of M holds then for some $j \leqslant n(\eta)$ we'll have $\Delta_{\eta,j}(W) =^* G_j$.

η is allowed to extend the definition of $\Delta_{\eta,j}(W)$ (i.e. to enumerate a new axiom into $\Delta_{\eta,j}$) at stages which are η-expansionary. As is implies by the intention, η always defines the value of $\Delta_{\eta,j}(W,x)[s]$ to be $G_j(x)[s]$.

η makes $\Delta_{\eta,j}$ monotone, so that at any stage we'll have $\operatorname{dom}\Delta_{\eta,j}(W)$ an ordinal. If η wishes to extend $\Delta_{\eta,j}$ at s then it defines one new axiom on the input $x = \operatorname{dom}\Delta_{\eta,j}(W)\,[s]$ (we'll ensure that the use is larger than the uses $\delta_{\eta,j}(W,y)\,[s]$ for all $y < x$ so that indeed monotonicity is maintained.)

To each computation $(\sigma; x, l) \in \Delta_{\eta,j}$ is associated a *chit* (y,π) (we sometimes also refer to y as the chit). When η wishes to define $\Delta_{\eta,0}(x)$ at stage s, it picks a new *suitable* chit. A chit (y,π) is suitable to be picked for a new computation $\Delta_{\eta,0}(W,x)\,[s]$ if:

(1) $y \in \alpha^{[\eta]}$.
(2) $y < \ell(M)[s]$.
(3) $\phi(W \oplus P, y)[s] = \sigma \oplus \pi$.
(4) $y > \mathtt{init}(\eta)[s]$ and $y > t$ for any $t < s$ at which η defined any $\Delta_{\eta,0}$ computation.

If s is η-expansionary and there is a suitable chit y, then it defines $\Delta_{\eta,0}(W,x)$ with use σ.

Now let $j > 0$. Suppose that at s, η wishes to define $\Delta_{\eta,j}(W,x)$. To do so, it needs to find a chit (y,π) which is suitable to be picked for this computation (we also say that the chit is *j-suitable*). The suitability conditions are:

(1) (y,π) is a chit for a computation $\Delta_{\eta,j-1}(W,x')[s]$ (whose use is σ).
(2) Further, that chit is still *active*, which means that $\pi \subset P\,[s]$ (an inactive chit is also called *cancelled*).
(3) The computation $\Delta_{\eta,j-1}(W,x')[s]$ is *failed*, which means that its value disagrees with $G_{j-1}(x')$ (necessarily x' entered G_{j-1} at some stage after the computation was defined).
(4) The size condition: $y > t$ for any $t < s$ at which η defined any $\Delta_{\eta,j}$ computation.

If such a chit is found then η defines $\Delta_{\eta,j}(W,x)[s]$ with the same use σ as the accompanying computation $\Delta_{\eta,j-1}(W,x')[s]$.

If a computation $\Delta_{\eta,j}(W,x)$ which is defined at s becomes incorrect at a later stage t (that is, some number smaller than the use enters W) then the accompanying chit (y,π) is cancelled and never considered again. Note again that if (y,π) is a chit for $\Delta_{\eta,j}(W, x_j)$ for $j \leqslant i$ (where $i \leqslant n(\eta)$) then the uses of all of these computations are the same and so all such computations disappear together). However, (2) shows that the chit may be cancelled even if the computations still hold.

At stage s, η may wish to use a chit (y, π) for purposes of victory. Suppose that $\rho \subset \eta$ is an agent which works for T_n at s. A chit (y, π) is *cleared* by ρ at stage s if it is *not* the case that

$$\Gamma_\rho(B \oplus G_n \oplus P, y) \downarrow = 0[s],$$

or if it is the case that $\Gamma_\rho(B \oplus G_n \oplus P, y) \downarrow = 0[s]$ but with use

$$\gamma_\rho(y)[s] > \operatorname{dom} \pi.$$

The chit y is *victorious* if it is M-confirmed at s, still active, is greater than $\texttt{Rest}(\eta)\,[s]$, and is cleared by all such agents ρ below η.

2.2.7. *Guessing.* This pertains only to type 2 constructions. The fact that $V_{i,j}$ are uniformly low implies that there are recursive functions f, g with the following properties: for all i, j, Ψ, η, $\lim_{s\to\alpha} f(Z_{i,j,\Psi}, s)$ and $\lim_{s\to\alpha} g(\eta, Z_{i,j,\Psi}, s)$ exist and are either $\texttt{yes}$ or $\texttt{no}$. Also,

- $\lim_{s\to\alpha} f(Z_{i,j,\Psi}, s) = \texttt{yes}$ iff there is some stage s of the construction and some $x \in G_i[s]$ which at s is realized ($\Psi(V_{i,j}, x) \downarrow = 0$) by a correct computation;
- $\lim_{s\to\alpha} g(\eta, Z_{i,j,\Psi}, s) = \texttt{yes}$ iff there is some stage s of the construction at which η works for $Z_{i,j,\Psi}$, η is eligible to act and does not believe it is satisfied, and η has a follower x which is permitted and is realized by a correct computation.

The recursion theorem allows us to use these function during the construction.

Suppose that at stage s, η is an agent working for $Z_{i,j,\Psi}$. Suppose that $x \in G_i[s]$ is realized at s. To check whether η believes this realization is correct, it looks for the least stage $t > s$ at which either the computation realizing x at s is discovered to be incorrect, or $f(Z, t) = \texttt{yes}$. If it gets the latter outcome then η believes the realization and believes that it is satisfied and does not need to act.

Suppose that η is eligible to act at s and that at s, η has a follower x which is realized and permitted. Before acting, η makes a guess about x's realization in much the same way: it looks for $t > s$ at which this realization is discovered to be incorrect, or $g(\eta, Z, t) = \texttt{yes}$. Again η believes this realization if the latter is the outcome.

2.3. Construction. The structure of every stage in all constructions is fixed. Every stage consists of two *pahses*: at the first, agents who have finitary-type actions to perform may do so (these actions need to be performed immediately, and the agents cannot wait until they are accessible.) At the second phase, less pressing matters are dealt with.

2.3.1. *The First Phase.* At the first phase, agents working for finitary-type requirements – D, N, Z, K requirements – that wish to act and are eligible to act, are allowed to act. An agent η is eligible to act at the first phase of stage s if there was a stage $t < s$ such that η was accessible at t, and such that between t and s, neither was η initialized, nor did it act.

No agent η wishes to act at s (at either phase) if it declared victory since $\texttt{init}(\eta)[s]$.

K. Suppose that at stage s, η works for $K_{\Xi,K}$. η wishes to act at s if

$$i(\eta)[s] \in \Xi(R \oplus P \oplus Q \oplus L)\,[s].$$

To act, it initializes weaker agents. As described above, after acting η either updates $i(\eta)$ or declares victory.

N and D. Suppose that at s, η works for some N or D requirement. The agent η may have three reasons to wish to act:

(1) It has a follower which is both realized and permitted.
(2) It has a follower which is realized for the first time.
(3) All of its followers are realized (this includes the possibility that there are no followers).

η's action corresponds to the case of affairs:

(1) η enumerates the follower into the sets it is targeted for and declares victory.
(2) η initializes weaker agents.
(3) η appoints a new, large follower (taken from $\alpha^{[\eta]}$), and initializes weaker agents.

Z. Suppose that at s, η works for some Z requirement. The agent η first guesses if it is already satisfied. If not, and if η has some follower that is both permitted and realized, then η guesses whether this realization is correct; if so, it wishes to act and indeed enumerates the follower into the sets for which it is targeted. It also initializes weaker agents.

If it didn't act, then η may yet wish to act if all of its followers are realized at s (it does not perform further guesses). In this case it appoints a new, large follower and initializes weaker agents.

We need to note something important. It may seem futile to let every agent that wishes to act at the first phase the right to do so, as action by a stronger requirement will immediately initialize it. This is correct, except for agents working for Z requirements. They may succeed even if initialized, and for the tactic of low guessing to succeed we need to allow them to act whenever they wish. Furthermore, the priority ordering on all agents may not be a well-ordering, so there isn't necessarily a strongest agent wishing to act. [This leads to intriguing consequences. Suppose we wished to overcome the first difficulty by adding the phrase "and no stronger requirement wishes to act" to the properties of g (and thus of the guessing). Suppose further that there is an infinite descending chain of agents for Z requirements which have realized, permitted followers ready for the testing. The agents now have no idea whether they should perform the guess; if the guess for a stronger agent succeeds then their own guess may never terminate.]

2.3.2. *The Second Phase.* At the second phase, all agents which are currently accessible and which work for an M requirement or a T requirement follow the following orders. We note that $\delta\,[s]$ *is* well-ordered, so if some agent wishes to declare victory and initialize weaker agents then we let the strongest one do so.

M. Suppose that at stage s, η works for $M = M_{\Theta,\Phi,W}$. As usual, if η declared victory since $\mathtt{init}(\eta)[s]$ then it does absolutely nothing.

If there is some $\sigma \subset \Theta(R)[s]$ and some $x < \operatorname{dom}\sigma$ such that $\sigma(x) = 0$ and $x \in W[s]$ then η declares victory (this is called "easy victory").

If easy victory was not declared, and if $r(\eta)[s] > 0$, then η does nothing at this stage. Otherwise, η looks for victorious chits. If there is one, it enumerates the least one y into Q, and for all $\rho \subset \eta$, if $\Gamma_\rho(G_{n(\rho)} \oplus P, y) \downarrow= 0[s]$, it puts $\gamma_\rho(G_{n(\rho)} \oplus P, y)[s]$ into P; η declares victory.

If η hasn't won yet, and the stage is η-expansionary, then η tries to extend $\Delta_{\eta,j}(W)$ as described above. Suppose η just defined $\Delta_{\eta,j}(x)$. Suppose that $\rho \supset \eta$ is an agent extending η which works for some D, Z or N requirement and has a follower x, targeted for G_j. Then ρ now initializes all weaker agents.

T. If $\eta \in \delta\,[s]$ works for T_n at s, then η extends Γ_η: it finds the least x such that $\Gamma_\eta(G_n \oplus P, x) \uparrow [s]$, and sets $\Gamma_\eta(G_n \oplus P, x) \downarrow= Q(x)\,[s]$ with large use.

2.3.3. *What we didn't specify.* We note that in order to specify a construction, all we need to do is specify what the tree of agents is, how to calculate $\delta\,[s]$ and how to assign requirements to agents at each stage; and we need to verify that $n(\rho) < \omega$ for every agent ρ.

2.4. Verifications. We first remark that the properties of the functions f and g ensure that every particular search must terminate by a bounded amount of time; either the realization of the number x being tested is incorrect, or x and s are witnesses to the limit value being `yes`.

It follows that the searching period at the beginning of the first phase of every stage does indeed take only α-finitely much time. For the function taking each instance of guessing to the stage at which the guess is resolved is recursive, and at each stage only α-finitely many guesses are made.

2.4.1. *Fairness Attempted.* Suppose that η is an agent that eventually stops being initialized:

$$r^* = \mathtt{init}(\eta)[\alpha] < \alpha.$$

It follows that η works for a fixed requirement after r^* (or works for no requirement at all). Further assume that η is accessible unboundedly often.

We examine how η affects weaker requirements.

M. If η works for an M requirement then η initializes weaker agents at most once after r^* (when declaring victory).

K. Suppose that η works for the requirement $K_{\Xi,K}$. After r^*, $i(\eta)[s]$ is non-decreasing. Let $K' = \{i(\eta)[s] : s > r^*\}$; K' is α-finite as it is an initial segment of K. The function which takes $x \in K'$ to the least $s > r^*$ at which $i(\eta)[s] = x$ is recursive and so bounded by some s^*. After s^*, η does not initialize weaker agents.

Suppose that η works for a D or an N requirement.

OBSERVATION 2.7. If x is a follower for η which is realized at $s > r^*$, then this realization is preserved by η's initializing of weaker agents at s.

LEMMA 2.8. *Eventually, η stops appointing new followers.*

PROOF. This is a standard permitting argument. If η declares victory at $s > r^*$ then after s, η ceases all action. Otherwise: suppose we are dealing with a type 1 construction. If at stage $s > r^*$, x is a follower for η which is realized, then $A \restriction x\,[s] = A \restriction x$. This is because x remains realized and η does not act; so x is never permitted after s. If the lemma fails then η keeps appointing followers which are unbounded in size (as it appoints large followers), and each follower eventually gets realized (by the time a larger follower gets appointed). This allows us to compute A.

If this is a type 2 construction and η appoints followers targeted for G_i, $i \in H$, then this argument holds, with U_i in place of A. If η does not target followers for G_i for any $i \in H$, then η only appoints one follower after r^*: whenever a follower gets realized it can be enumerated. $\square$

Suppose that η works for $Z = Z_{i,j,\Psi}$.

LEMMA 2.9. *η eventually stops appointing new followers or enumerating followers into G_i.*

PROOF. This is classical. If $\lim f(Z,s) = \texttt{yes}$, then from the stage at which $f(Z,s)$ stabilizes, η always guesses itself to be successful and does nothing. If $\lim g(\eta, Z, s) = \texttt{yes}$ then $\lim f(Z,s) = \texttt{yes}$; the witness for the former is believed useful and enumerated.

Suppose that $\lim g(\eta, Z, s) = \lim f(Z,s) = \texttt{no}$ and both stabilize after $t > r^*$. Then after t, no follower is deemed useful, so η doesn't enumerate followers after t.

Suppose that η keeps appointing new followers. Let x be a follower for η, appointed at $s > r^*$. At every stage $u > s$ at which η appoints another follower, x is realized. The enumeration $\langle V_{i,j}[s]\rangle$ is low for Ψ so x is eventually correctly realized. After t^*, a correctly realized follower cannot be permitted: η does not believe itself successful after t^* and so a correctly realized, permitted follower would be a witness for $\lim_{s\to\alpha} g(\eta, Z, s) = \texttt{yes}$. As the size of followers appointed is unbounded, this gives us a procedure of computing U_i from $V_{i,j}$. $\square$

It follows that if η works for a D, N or Z requirement, then η eventually stops initializing weaker requirements *on its own accord.* We still have to verify it stops initializing weaker requirements when it is instructed to do so by a stronger agent. Suppose then that $\rho \subsetneq \eta$ works for $M = M_{\Theta,\Phi,W}$, fix $j \leqslant n(\rho)$ and suppose that η targets followers for G_j.

OBSERVATION 2.10. Suppose that $s > t > \texttt{init}(\rho)[\alpha]$. Suppose that $\Delta_{\rho,j}(W,x)\downarrow$ $[t]$ with use σ, and that $\sigma \subset \Theta(R)[s]$. Then ρ does not redefine $\Delta_{\rho,j}(x)$ at stage s. For if it does, then $\Delta_{\rho,j}(W,x)\uparrow$ $[s]$ which means that $\sigma \not\subset W[s]$ anymore; necessarily some $y < \operatorname{dom}\sigma$ entered W between t and s. But then ρ can get an easy victory at s and defines nothing.

Let $K(\eta)$ be the final set of followers for η (it is α-finite), and let $r^{**} > r^*$ be a stage after which η neither appoints any followers, nor enumerates followers into sets. Suppose that $x \in K(\eta)$ and suppose that at some stage $s > r^{**}$, ρ defines $\Delta_{\rho,j}(x)$, say with use σ. At some stage $t > r^*$, x is chosen as a follower for η. Since large followers are chosen, we must have $t < s$; so η initializes weaker agents at s. η's action at s ensures that $\sigma \subset \Theta(R)$ from s onwards (this is because $s > r^{**}$). It follows that ρ will not redefine $\Delta_{\rho,j}(x)$ at any other stage after r^{**}.

Let $K'(\eta,\rho,j)$ be the set of followers $x \in K(\eta)$ such that η initializes for the sake of a $\Delta_{\rho,j}(x)$ computation after stage r^{**}. Let $x_0 = \min K'(\eta,\rho,j)$, $x_2 \in K'(\eta,\rho,j)$ and $x_1 \in K(\eta)\cap(x_0,x_1)$. Say that η initializes on behalf of $\Delta_{\rho,j}(x_0)$ at $s_0 > r^{**}$ and on behalf of $\Delta_{\rho,j}(x_2)$ at $s_2 > r^{**}$. By the above analysis, and since $\Delta_{\rho,j}(W,x_0)\downarrow$ $[s_2]$, we know that $s_2 > s_0$ and that at some stage $s_1 \in (s_0,s_2)$, ρ defines $\Delta_{\rho,j}(x_1)$. It follows that $x_1 \in K'(\eta,\rho,j)$. Thus $K'(\eta,\rho,j)$ is the intersection of $K(\eta)$ with an interval, and so is α-finite.

For $x \in K'(\eta, \rho, j)$, let $s_{\eta,\rho,j}(x)$ be the (unique) stage at which ρ defines $\Delta_{\rho,j}(x)$ (and η initializes for x). As $K'(\eta, \rho, j)$ is α-finite, $s_{\eta,\rho,j}$"$K'(\eta, \rho, j)$ is bounded. We get:

LEMMA 2.11. *Let η be any agent such that $\texttt{init}(\eta)[\alpha] < \alpha$. There is a stage after which η does not initialize weaker requirements on its own behalf. If η works (after $\texttt{init}(\eta)[\alpha]$) for a D, N or a Z requirement, then for every $\rho \subset \eta$ there is a stage after which η does not initialize on ρ's behalf.*

2.4.2. *The Fairness Assumption.* We now make the fairness assumption on the construction:

> For every requirement R there is an agent η which is not initialized unboundedly often ($\texttt{init}(\eta)[\alpha] < \alpha$), is accessible unboundedly often, and which eventually works for R. Furthermore, if R is an M requirement, then there are club many stages at which η is accessible and on which $\texttt{Rest}(\eta)$ is bounded.

We show that if the assumption holds then the construction succeeds.

2.4.3. *Success of the Finitary-Type Requirements.*

LEMMA 2.12. *Every K requirement is met.*

PROOF. Suppose that η is the agent working for $K_{\Xi,K}$. If η acts at $s > \texttt{init}(\eta)[\alpha]$ then $i(\eta)[s] \in \Xi(R \oplus P \oplus Q \oplus L)$ is permanently correct. Thus if η declares victory after $\texttt{init}(\eta)[\alpha]$ then it succeeds.

If not, let i be the final value of $i(\eta)[s]$, stabilizing after s^*. For no $s > s^*$ do we have $i \in \Xi(R \oplus P \oplus Q \oplus L)\,[s]$ and so the requirement is met vacuously. □

COROLLARY 2.13. *All sets constructed are amenable.*

PROOF. Fix $\beta < \alpha$. Let $\Xi = \{p : \operatorname{dom} p > \beta\}$. Success of K_Ξ implies that there is some stage $s < \alpha$ after which $R \oplus P \oplus Q \oplus L \restriction \beta$ is fixed. □

COROLLARY 2.14. *All sets constructed are low, hence admissible.*

PROOF. We show that some functional is low for $X = R \oplus P \oplus Q \oplus L$. Fix some Ξ such that $\Xi(X) = X'$. Success of $K_{\Xi,\{x\}}$ for all $x < \alpha$ shows that Ξ is weakly low for X. Suppose further that K is α-finite; we want to show that if $K \subset X'$ then from some stage we have $K \subset \Xi(X)\,[s]$. If K has a maximal element then this is ensured by the success of $K_{\Xi,K}$. Otherwise, this is ensured by the success of $K_{\Xi',K'}$ where $K' = K \cup \{\sup K\}$ and $\Xi' = \Xi \cup \{(\langle\rangle, \sup K)\}$. □

Let η be the agent working for T_n.

CLAIM 2.15. *If $\Gamma_\eta(G_n \oplus P, y)\downarrow$ then $\Gamma_\eta(G_n \oplus P, y) = Q(y)$.*

PROOF. Let $s > r^* = \texttt{init}(\eta)[\alpha]$, and suppose that $\Gamma_\eta(G_n \oplus P, y)\downarrow [s]$ and that at s, agent ρ, working for $M_{\Theta,\Phi,W}$ puts y into Q. ρ cannot be stronger than η since $s > r^*$. Also, ρ cannot be to the right of η; η defined $\Gamma_\eta(y)$ at stage $t < s$ and initialized all nodes to its right at t; and new chits y are picked large. Thus $\eta \subset \rho$.

Therefore ρ puts $\gamma_n(y)[s]$ into P at s, removing the computation $\Gamma_\rho(G_n \oplus P, y)$. If the computation is defined later then the value must be correct. □

CLAIM 2.16. *Suppose that $\beta < \alpha$ and that $\Gamma_\eta(G_n \oplus P) \restriction \beta$ stabilizes by some α-finite stage. Then $\Gamma_\eta(G_n \oplus P, \beta)\downarrow$.*

PROOF. Suppose that after $t > r^*$, $\Gamma_\eta(G_n \oplus P) \restriction \beta$ is fixed. At every later stage at which η is allowed to define Γ_ρ, if $\Gamma_\rho(G_n \oplus P, \beta) \uparrow$ then it gets defined, say with use τ. At such a stage enumerate τ into a functional Ξ. Let ρ be the agent which works for K_Ξ. At some stage $t > s$, Ξ acts and so preserves $\tau \subset G_n \oplus P$. □

LEMMA 2.17. *Every T requirement is met.*

PROOF. For every β, $\Gamma_\eta(G_n \oplus P) \restriction \beta$ eventually stabilizes. This follows from the fact that $G_n \oplus P$ is admissible and is proved by induction. Assume up to β; the function taking $y < x$ to the stage at which the correct computation $\Gamma_\eta(G_n \oplus P, y)$ is defined is weakly recursive in $G_n \oplus P$ and so is bounded.

It follows that $\Gamma_\eta(G_n \oplus P) = Q$. As $G_n \oplus P$ is admissible, $Q \leqslant_\alpha G_n \oplus P$. □

Let η be the agent responsible for a D or an N requirement, and let $K(\eta)[s]$ be the set of followers η has at stage $s > r^*$ ($K(\eta) = K(\eta)[\alpha]$).

CLAIM 2.18. *Every $x \in K(\eta)[s]$ is realized at s, except perhaps for a maximal element.*

PROOF. We show this by induction on s. Assume that s is a limit stage and that $x \in K(\eta)[s]$ is not maximal. Take some $y > x$, $y \in K(\eta)[s]$. For some $t < s$, $x, y \in K(\eta)[t]$. By induction, x is realized at t. By observation 2.7, x is realized at s as well.

Assume the statement holds for $K(\eta)[s]$. Then it holds for $K(n)[s+1]$, because at stage s all previously realized numbers are still realized (same observation), and η perhaps appoints a new follower, larger than $\sup K(\eta)[s]$, but only if every $x \in K(\eta)[s]$ is realized. □

LEMMA 2.19. *Every D and N requirement succeeds.*

PROOF. If η declares victory after $r^* = \texttt{init}(\eta)[\alpha]$ then the diagonalization is preserved. Otherwise, suppose that after $t > r^*$, η appoints no new followers. This means that there is one maximal follower x which is never realized. Success follows. □

LEMMA 2.20. *Every Z requirement succeeds.*

PROOF. Let η be the agent working for $Z = Z_{i,j,\Psi}$. If $\lim f(Z, s) = 1$ then success is clear. If not, let $K(\eta)$ be the set of final followers (that are not enumerated into G_i); after some $t > \texttt{init}(\eta)[\alpha]$, η does not enumerate any followers or appoint new ones.

We claim that there is some $x \in K(\eta)$ which is not realized (at α). Otherwise, there is a stage at which all permanent followers are correctly realized: the function taking the follower to the stage at which it is correctly realized is computable from $V_{i,j}$, which is admissible, and $K(\eta)$ is α-finite. But after that stage, η would appoint a new follower: it does not believe itself to be successful. □

Suppose that $j \preccurlyeq i$. We want to show that $G_j \leqslant_{w\alpha} G_i \oplus L$. To determine whether $x \in G_j$, we first wait until stage x to see if x was chosen as a follower for some agent η which targets x to G_j. If not then $x \notin G_j$. Suppose that it is. If η works for $D_{k,j,\Psi}$ or $Z_{j,k,\Psi}$ then $x \in G_j$ iff $x \in L$. If η works for $N_{k,l,\Psi}$ (so $l \preccurlyeq j$)

then $x \in G_j$ iff $x \in G_i$ (since $l \preccurlyeq i$ too). Admissibility implies that $G_j \leqslant_\alpha G_i \oplus L$. Overall we see that indeed the partial ordering coded by L is $\preccurlyeq$.

Finally we mention that the permitting method indeed ensures that in type 1 constructions, $R, L \leqslant_\alpha A$, and that in type 2 constructions, $G_i \leqslant_\alpha U_i$ for $i \in H$.

2.4.4. *The Minimality Requirements.* Let η be the agent working for $M = M_{\Theta,\Phi,W}$, and let $r^* = \mathtt{init}(\eta)[\alpha]$.

OBSERVATION 2.21. Suppose that $s > t > r^*$. Suppose that at t, (y, π) is picked as a chit for a computation $\Delta_{\eta,0}(x)$ with use σ. Suppose that at s, (y, π) is transferred as a chit for a $\Delta_{\eta,j}$ computation. Definitions of Δ computations are done only at η-expansionary stages, so $y < l(M)[t] < l(M)[s]$, so y is M-confirmed at s. Also, (y, π) is still active at s. This implies that $\phi(W \oplus P, y)[s] = \sigma \oplus \pi$ (we use niceness of Φ here). Thus $\sigma \subset \Theta(R)[s]$. Note that once the chit is cancelled it is never active again (and will never be picked again as a new chit for $\Delta_{\eta,0}$ as we pick large chits).

LEMMA 2.22. *If η declares victory after r^*, then the hypotheses of M do not hold.*

PROOF. If easy victory is declared then η preserves a discrepancy between $\Theta(R)$ and W. Suppose that at $s > r^*$, η enumerates a victorious chit y into Q. Let $\sigma \oplus \pi = \phi(W \oplus P, y)[s]$. $\sigma \subset \Theta(R)\,[s]$ holds by observation 2.21 and is preserved by η's action at s (so if $\sigma \not\subset W$ we again get an easy win.) We claim that $\pi \subset P$ is preserved (and so that y is not M-confirmed at α). The only obstacle can be the uses of the form $\gamma_\rho(G_n \oplus P, y)$ which η enumerates into P as it declares victory. A use of this kind is enumerated only if Γ_ρ clears y at s, thus all $\gamma_\rho(G_n \oplus P, y)$ enumerated are greater than $\operatorname{dom} \pi$. ☐

The following lemma will be needed here and also in the following section in order to ensure that the fairness assumption holds.

CLAIM 2.23. *$r(\eta)[s]$ is bounded on a recursive club.*

PROOF. $r(\eta)[s]$ is constant on a recursive club: if there are unboundedly many η-expansionary stages, then the set of stages at which $r(\eta) = 0$ is a (recursive) club. If not, then $r(\eta)[s]$ is eventually constant. ☐

Assume that the hypotheses of M hold.

CLAIM 2.24. $\lim_{s \to \alpha} l(M)\,[s] = \alpha$.

PROOF. This is because $W \oplus P \oplus R$ is admissible. The least stage at which a number x is M-confirmed with correct uses for $\Theta(R)$ and $\Phi(W \oplus P)$ is computable (as a function of x) from $W \oplus P \oplus R$; hence bounded on initial segments of α. ☐

It follows that there are unboundedly many M-expansionary stages.

Let $n = n(\eta)$.

CLAIM 2.25. *Let $j \leqslant n$. For all x, there is a stage after which η stops defining $\Delta_{\eta,j}(x)$.*

PROOF. At each stage s at which η defines $\Delta_{\eta,j}(x)$, say with use σ, enumerate $\theta(R; \operatorname{dom}\sigma)[s]$ into a functional Ξ (recall observation 2.21). The success of the agent ρ which works for K_Ξ ensures that if unboundedly many attempts at defining $\Delta_{\eta,j}(x)$ are made, then $\theta(R; \operatorname{dom}\sigma)[s]$ for some such definition will be preserved. By observation 2.10, $\Delta_{\eta,j}(x)$ doesn't get redefined after s. □

CLAIM 2.26. $\operatorname{dom}\Delta_{\eta,0}(W) = \alpha$.

PROOF. By induction on x, we show that $\Delta_{\eta,0}(W,x)\downarrow$. If $x \leqslant \operatorname{dom}\Delta_{\eta,0}(W)$, then by admissibility of W, we know that $\Delta_{\eta,0}(W) \restriction x$ eventually stabilizes. It is enough now to show that for unboundedly many s, $\Delta_{\eta,0}(W,x)\downarrow[s]$; by claim 2.25, after some t, η stops defining $\Delta_{\eta,0}(x)$, hence all computations $\Delta_{\eta,0}(W,x)[s]$ for $s > t$ must be the same computation, which is permanent.

Suppose that by $s^* > r^*$, $\Delta_{\eta,0}(W) \restriction x$ is permanent. Suppose that $t > s^*$ and $\Delta_{\eta,0}(W,x)\uparrow[t]$. Find some M-expansionary stage $s > t$ such that $l(M)[s]$ is large enough so that there is some $y \in \alpha^{[\eta]}$ such that $t < y < l(M)[s]$. Now if η didn't define $\Delta_{\eta,0}(x)$ between t and s, then y is a suitable chit (because η did not define any $\Delta_{\eta,0}$ computations between t and s), thus η would define $\Delta_{\eta,0}(x)$ at s. □

CLAIM 2.27. *Say $j \in [1,n]$. If there are unboundedly many chits which are eventually $j-1$-suitable (and never cancelled), then* $\operatorname{dom}\Delta_{\eta,j}(W) = \alpha$.

PROOF. This is like the previous claim. Letting x, s^*, t be as above, we find an expansionary stage $s > t$ such that at s, there is a chit y which is already $j-1$-suitable and is never cancelled, such that $t < y < l(M)[s]$; this is possible by the assumption that there are unboundedly many such y. □

CLAIM 2.28. *Suppose that at stage s, y is an active chit for a failed $\Delta_{\eta,j}$ computation. Suppose that $\rho \subset \eta$ works for T_j. Then Γ_ρ clears y at s.*

PROOF. We show that if $\Gamma_\rho(G_j \oplus P, y)\downarrow[s]$ then this computation must have been defined after the stage $t < s$ at which the chit was originally picked as a chit for a computation $\Delta_{\eta,j}(x)$. Obviously (by the size requirement for y) we have $x \leqslant y$. As the computation is failed at s, x enters G_j at a stage $u \in (t,s)$; now $x \leqslant y \leqslant \gamma_\rho(G_j \oplus P, y)[u]$ (if $\Gamma_\rho(G_j \oplus P, y)\downarrow[u]$), destroying the latter computation. □

LEMMA 2.29. *There cannot be unboundedly many chits which are eventually n-suitable and which are never cancelled.*

PROOF. We show that if there are unboundedly many such chits then η declares victory after r^*.

By the fairness assumption, there is some $\beta < \alpha$ such that

$$C_0 = \{s > r^*,: \texttt{Rest}(\eta)[s] \leqslant \beta \ \& \ \eta \in \delta[s]\}$$

contains a recursive club. Further, we know that

$$C_1 = \{s > r^* : r(\eta)[s] = 0\}$$

is a recursive club. Let $C = C_0 \cap C_1$; C contains a recursive club.

By assumption we can find $y > \beta$ and a stage $s \in C$ such that at s, y is an active chit for a failed $\Delta_{\eta,n}$ computation and such that $y < l(M)[s]$. Then y is victorious at s, and at s, η may win by enumerating y into Q. □

The following concludes the verifications. For $A, B \subset \alpha$, we say that $A =^* B$ if there is some $\beta < \alpha$ such that $A \upharpoonright (\beta, \alpha) = B \upharpoonright (\beta, \alpha)$. A corollary of the next and previous lemmas is that there is some $j \leqslant n$ such that $\Delta_{\eta,j}(W) =^* G_j$. As W is admissible and G_j amenable we get $G_j \leqslant_\alpha W$.

LEMMA 2.30. *Suppose that $j \leqslant n$ and $\operatorname{dom}\Delta_{\eta,j}(W) = \alpha$ but $\Delta_{\eta,j}(W) \neq^* G_j$. Then there are unboundedly many chits which are eventually j-suitable and are never cancelled.*

PROOF. Much of the proof goes along classical lines. Fix $\beta < \alpha$. Let the functional Ξ converge at $t > r^*$ with use $\rho \oplus \pi \subset R \oplus P\,[t]$ if there is some $\sigma \subset W[t]$ such that $\sigma \subset \Theta(\rho)[t]$ and there is some $y > \beta$ such that at t, (y, π) is an active chit for a failed $\Delta_{\eta,j}$ computation with use σ. If Ξ converges unboundedly often, the success of K_Ξ would show that there is some j-suitable chit which is never cancelled and which is greater than β (protection of $\rho \oplus \pi \subset R \oplus P$ ensures that $\sigma \subset W$).

Suppose that $t^* > r^*$ is any stage. Take some $x > \beta, t^*$ such that $\Delta_{\eta,j}(W, x) \neq Q(x)$. Suppose that the correct $\Delta_{\eta,j}(x)$ computation is defined at stage $s > t^*$ with associated chit (y, π). Let $t > s$ be the stage at which this computation fails; some agent ρ enumerates x into G_j at t. The standard argument shows that $\eta \subset \rho$: ρ cannot be stronger than η as $t > r^*$, and η cannot lie to the right of ρ since η initializes nodes to its right at s and $s > x$. This argument shows that x is a follower for ρ at s and so that ρ initializes weaker requirements on η's behalf at s.

Let u be the least stage greater than t at which $r(\eta) = 0$. The key is noticing that u is M-expansionary, as it cannot be a limit of M-expansionary stages. Thus $y < l(M)[u]$.

We verify that $\pi \subset P$ is preserved until u. As mentioned, ρ initializes weaker agents at s. ρ is not initialized until at least after t, as x is still a follower at t. Thus $\pi \subset P$ is preserved between s and t by ρ's action. Between t and u, restraint which is imposed by η protects $\pi \subset P$; recall that s is an M-expansionary stage.

Now at u, y is M-confirmed so $\sigma \subset \Theta(R)[u]$ (note that $\sigma \subset W$ is always true since we picked the correct $\Delta_{\eta,j}(W)$ computation). Thus $\Xi(R \oplus P) \downarrow [u]$. □

3. Various Constructions

Here we apply the template to particular classes of admissible ordinals α. We recall that we need to describe the tree of agents, assign requirements, define $\delta\,[s]$, and make sure that $n(\rho) < \omega$ for all ρ and that the fairness assumption holds.

3.1. $\varrho_\alpha = \omega$.

Suppose that $\varrho_\alpha = \omega$. Fix a partial recursive, 1-1 and onto $p \colon \omega \twoheadrightarrow \alpha$.

The biggest obvious difference between the classical construction and this one is the number of requirements. Initially, we arrange the requirements effectively in order-type α. We then use the map p in order to re-arrange the requirements in order-type ω. The tree of agents is thus ω, and at stage s, agent m is assigned the requirement $p(m)\,[s]$; if $p(m) \uparrow [s]$ then m is not assigned any requirement at stage s. We let $\delta\,[s] = \omega$ for all s. As every agent is preceded by only finitely many agents, $n(m)[s] < \omega$ for all m and s.

The fairness assumption is not hard to verify. Each agent is assigned a requirement at most once. By induction we show for every agent m that $\texttt{init}(m)[\alpha] < \alpha$ and that eventually m does not initialize weaker agents. The first holds by induction, because agent m is initialized only when it is assigned a requirement or when stronger agents initialize it. The second holds because of lemma 2.11 and the fact that for every m there are only finitely many agents $k < m$.

The fact that $\alpha > \omega$ and as far as we can effectively tell, is a regular cardinal, allows for some simplifications in comparison with the classical construction: we no longer require a tree of outcomes to guess the true restraint on a particular agent. For all k, $r(k)$ is constant on a recursive club C_k (claim 2.23) and so $\texttt{Rest}(m)$ is bounded on $\bigcap_{k<m} C_k$, which is a recursive club.

[We notice that the price to pay for this simplification is that the restraint has to fall back on limits of expansionary stages, which may be not expansionary themselves; thus in theory a requirement might wish to impose further restraint at such a limit stage, and dropping the restraint might damage the success of the requirement. We noticed however that this does not happen in our construction, as the responsibility to act often lies with the lowness requirements associated with specific tasks (which are of a simpler, finitary nature); the minimality (M) requirements only need to act at successor expansionary stages, so dropping the restraint at limits of expansionary stages does not harm their success.]

3.1.1. *A Remark on the Case $\alpha = \omega_1^{CK}$.* As mentioned in section 1, if $\alpha = \omega_1^{CK}$ then Kleene's $\mathcal{O}$ is not α-recursive, so the theorems proved do not imply that models with Kleene's $\mathcal{O}$ can be coded. However, $\mathcal{O}$ is ω_1^{CK}-r.e., and so can be recursively approximated. We can modify both theorems 2.3 and 2.4 to make $\preccurlyeq$ recursively approximated in a Δ_2-fashion, rather than α-recursive. There are only minor changes to the construction. A requirement $N_{i,j,\Psi}$ may change its mind about its own necessity, but eventually it guesses correctly. As far as a D- or an N-requirement η initializing on the behalf of a stronger M-requirement ρ, to avoid confusion, we let the former initialize whenever the latter define a $\Delta_{\rho,j}(x)$ computation when x is a follower for η, no matter to which set it is targeted. It is easy to see this doesn't affect fairness of the construction. Finally, the positive ordering requirements $G_j \leqslant_\alpha G_i \oplus L$ for $j \preccurlyeq i$, also may change their mind, but eventually, every number put into j is put into L or G_i and the reduction still holds.

The full details can be found in [**GSS**].

3.2. $\varrho_\alpha^2 = \omega$ but $\varrho_\alpha > \omega$.

The main hurdle in this case is of course that we cannot effectively put the requirements in a list of length ω; we can only effectively approximate such a list. The problem with the approximation is that an agent which is not assigned any requirement may guess unboundedly often that it is assigned some requirement; under the previous rules, each such time would be destructive to weaker agents. [Thus if the Δ_2-projectum of α is ω then the previous construction can be performed without difficulty.]

To overcome this difficulty, we let the tree of agents be $2^{<\omega}$. Each agent guesses whether it, and the agents below it, would be really assigned a requirement or not; agents whose guess is correct will perform their duty successfully.

We noticed that a pure tree construction interferes with the finitary-type requirements, which demand immediate access when they discover a computation they wish to protect, or a follower that is permitted; they cannot wait. We thus rigged the construction in their favor by introducing the first phase of each stage, at which agents working for finitary requirements are allowed to elbow their way to the front of the line. As we shall shortly see, this implies that in order to prove fairness, we need to use the assumption that $\varrho_\alpha > \omega$. Thus the arguments for the case $\varrho_\alpha = \omega$ are actually needed independently.

We need to approximate a partial, onto map $\omega \to \alpha$; for this we need a notion of approximation of a partial function. We use the following:

DEFINITION. Let $\beta \leqslant \alpha$ and $f\colon \beta \to \alpha$ be partial. A function $f(x)[s]$ is a *strong approximation* of f if:

(1) For every $x \in \operatorname{dom} f$, $f(x)[s] = f(x)$ on all s but an initial segment of α.
(2) For every $x \notin \operatorname{dom} f$, $f(x)[s] = 0$ for unboundedly many stages s.
(3) For all x, the set of stages s at which $f(x)[s] = 0$ is closed.

The approximation is *tame* if for all $\gamma < \beta$, (1-3) hold at once for all $x < \gamma$. Namely: for all $\gamma < \beta$ there is some s^* such that:

a. For all $x \in \gamma \cap \operatorname{dom} f$ and $s > s^*$, $f(x)[s] = f(x)$.
b. $K_\gamma = \{s : f(x)[s] = 0 \text{ for all } x \in \gamma \setminus \operatorname{dom} f\}$ is a club.

[It follows that if f has a recursive tame strong approximation, then for all $\gamma < \beta$, $\gamma \cap \operatorname{dom} f$ and $f \restriction (\gamma \cap \operatorname{dom} f)$ are α-finite. If f has a recursive strong approximation then we could in fact, in (b), only require that the set K_γ be merely unbounded, rather than a club; it would still follow that $\gamma \cap \operatorname{dom} f$ is α-finite, and then K_γ would be a uniform intersection of recursive clubs (the stages $s > s^*$ at which $f(x)[s] = 0$ for any particular $x \notin \operatorname{dom} f$) and so a club.]

If $\operatorname{dom} f = \omega$ then any recursive strong approximation of f is also tame; this is because a finite intersection of recursive clubs is a club.

PROPOSITION 2.31. *For every admissible α and $\beta \leqslant \alpha$, every $\Sigma_2(J_\alpha)$ $f\colon \beta \to \alpha$ has a recursive strong approximation.*

PROOF. We approximate f in a similar way to that of lemma B.1. Let Φ be a nice, weak functional such that $\Phi(0') = f$. For any $x < \beta$ and $s < \alpha$, if there is some $s^* < s$ and some i such that for all $t \in [s^*, s]$ we have $\Phi(0', x)[s] = i$ then let $f(x)[s] = i$. Otherwise, let $f(x)[s] = 0$. □

Fix $p\colon \omega \to \alpha$ which is partial, $\Sigma_2(J_\alpha)$, 1-1 and onto. Get a recursive, strong approximation $p\,[s]$. $\delta\,[s] \in 2^\omega$ is defined by $\delta\,[s] = 1$ iff $p(n)\,[s] > 0$.

As usual, all requirements are arranged effectively in a list of length α. Call an agent η *active* if $\eta(|\eta| - 1) = 1$. An active agent η is assigned to requirement $p(|\eta| - 1)\,[s]$ if the latter converges at s. Passive (that is, non-active) nodes are never assigned requirements. Again since every agent is preceded by finitely many agents, $n(\rho)\,[s] < \omega$ for all ρ and s.

The true path is, of course, $\operatorname{dom} p$. We let, for $m < \omega$,

$$C_m = \{s : \operatorname{dom} p \restriction m \subset \delta\,[s]\};$$

Since $p\,[s] \to p$ is a tame strong approximation, we know that C_m contains a club.

By induction we can show that for every m, $\texttt{init}(\operatorname{dom} p \restriction m)[\alpha] < \alpha$ and that $\operatorname{dom} p \restriction m$ does not initialize weaker requirements. Assume up to m. If $\operatorname{dom} p \restriction m$ is passive then it is never assigned any requirement. Otherwise $p(m)\downarrow$ correctly by some stage, and after that stage $\operatorname{dom} p \restriction m$ will not be initialized because of a reassignment of its requirement. By induction there is a stage after which no $\rho \subset \operatorname{dom} p \restriction m$ initialized $\operatorname{dom} p \restriction m$.

In both cases we know that there is some stage s_0 after which no ρ which lies to the left of $\operatorname{dom} p \restriction m$ is accessible. Let K be the set of agents which lie to the left of $\operatorname{dom} p \restriction m$ that act after stage s_0. This is an r.e. subset of $2^{<\omega}$. Of course $|2^{<\omega}|^{J_\alpha} = \omega$, and $\varrho > \omega$, so K is α-finite. Every $\rho \in K$ acts exactly once after s_0. The function taking $\rho \in K$ to the stage at which it acts is recursive, hence bounded. Thus eventually $\operatorname{dom} p \restriction m$ is not initialized by nodes to its right, and we have $\mathtt{init}(\operatorname{dom} p \restriction m)[\alpha] < \alpha$.

There are only finitely many $\rho \subset \operatorname{dom} p \restriction m$, so lemma 2.11 shows that eventually $\operatorname{dom} p \restriction m$ stops initializing nodes extending it.

By claim 2.23, for every k there is a recursive club D_k on which $r(\operatorname{dom} p \restriction k)$ is bounded. It follows that on the club $\bigcap_{k<m} D_k \cap C_m$, $\operatorname{dom} p \restriction m$ is accessible and $\mathtt{Rest}(\operatorname{dom} p \restriction m)$ is bounded.

3.3. α is Σ_2-admissible, but $\operatorname{cf}_{\Sigma_3(J_\alpha)}(\alpha) = \omega$. To enable us to handle trees of height greater than ω, we assume Σ_2-admissibility.

Let $\lambda = \varrho^2_\alpha$.

CLAIM 2.32. *Let $g\colon \lambda \to \alpha$ be $\Sigma_2(J_\alpha)$ and partial. Then for all $\beta < \lambda$, $g \restriction \beta$ is α-finite.*

PROOF. The structure $M = (J_\alpha, \in, 0')$ is admissible. $\lambda = \varrho_M$ and g is M-partial recursive. For every $\beta < \lambda$, $\beta \cap \operatorname{dom} g$ is M-r.e. and so α-finite, and so $g \restriction \beta$ is $\Delta_1(M)$ and so α-finite. □

It follows that every such g has a recursive, tame strong approximation. It has a recursive strong approximation by proposition 2.31. Further, for every $\beta < \lambda$, the function taking $x \in \beta \cap \operatorname{dom} g$ to the least s such that $g(x)[t]$ is constant on $[s, \alpha)$ is $\Pi_1(J_\alpha)$ on an α-finite set, hence bounded. Also, as $\beta \setminus \operatorname{dom} g$ is α-finite,

$$\{s < \alpha \;:\; \text{for all } x \in \beta \setminus \operatorname{dom} g,\; g(x)[s] = 0\}$$

is an α-finite intersection of recursive clubs, hence is a club. Thus every recursive, strong approximation of g is tame.

$\Sigma_3(J_\alpha)$ functions will not have strong approximations; we need a notion of a weak approximation.

DEFINITION. Let $\beta \leqslant \alpha$ and $f\colon \beta \to \alpha$ be total. $f\colon \beta \times \alpha \to \alpha$ is a *weak approximation* of f if for all $x < \beta$,

(1) $\{s : f(x)[s] = f(x)\}$ is unbounded.
(2) $f(x)[s] \geqslant f(x)$ except for s in an initial segment of α.
(3) For all limit stages t, $f(x)[t] \leqslant \liminf_{s\to t} f(x)[s]$.

It follows that for all x for some $s^* < \alpha$, $\{s > s^* : f(x)[s] = f(x)\}$ is a club.

Again, a *tame*, weak approximation is an approximation for which (1-3) hold simultaneously for all $\gamma < \beta$. Namely, for all $\gamma < \beta$ there is some $s^* < \alpha$ such that for all $x < \gamma$ and $s > s^*$, $f(x)[s] \geqslant f(x)$ and $\{s : f(x)[s] = f(x) \text{ for all } x < \gamma\}$ contains a club. If $\operatorname{dom} f = \omega$ then any weak approximation is also tame.

Let $p\colon \lambda \to \alpha$ be partial, $\Sigma_2(J_\alpha)$, 1-1 and onto, and let $p[s]$ be a recursive, tame, strong approximation to p.

CLAIM 2.33. $\operatorname{cf}_{\Sigma_3(J_\alpha)}(\lambda) = \omega$.

PROOF. Let $f\colon \omega \to \alpha$ be $\Sigma_3(J_\alpha)$ and cofinal. Consider $p^{-1} \circ f\colon \omega \to \lambda$; it is a $\Sigma_3(J_\alpha)$ function. Let $\beta = \sup \operatorname{range} p^{-1} \circ f$. If $\beta < \lambda$ then $p``\beta$ is bounded below $\alpha$, but $p``\beta$ contains $\operatorname{range} f$. □

LEMMA 2.34. *There is an increasing and cofinal $f\colon \omega \to \lambda$ that has a recursive, weak approximation $f[s]$ such that each $f[s]$ is increasing.*

PROOF. Fix $g\colon \omega \to \lambda$ cofinal such that

$$g(n) = \beta \Leftrightarrow \exists x \forall y \exists z\, R(n, \beta, x, y, z)$$

where R is recursive. For simplicity assume that always $x > 0$. Let the length of agreement be

$$l(n, \beta, x)[s] \;=\; \min\{y < s \;:\; \neg\exists x\, R(n, \beta, x, y, z)\}.$$

Use the standard pairing function $i \to ((i)_0, (i)_1)$; λ is closed under this pairing.

Let $f_0(n)[s]$ be the least $i < \lambda$ such that letting $\beta = (i)_0$ and $x = p((i)_1)[s] > 0$, for all $t < s$, $l(n, \beta, x)[t] < l(n, \beta, x)[s]$ (if no such pair exists let $g_0(n)[s] = s$.)

Fix $n < \omega$. Let $i(n)$ be the least $i < \lambda$ such that for $\beta = (i)_0 (= g(n))$ and $x = p((i)_1)$, for all $y < \alpha$ there is some $z < \alpha$ such that $R(n, \beta, x, y, z)$.

CLAIM 2.35. $\lim_{s \to \alpha} l(n, \beta, x)[s] = \alpha$.

PROOF. By admissibility; the function taking $y < \alpha$ to the least z such that $R(n, \beta, x, y, z)$ is recursive and so bounded on every $y < \alpha$. □

Let $K_n = \{i < i(n) : (i)_1 \in \operatorname{dom} p\}$; K_n is α-finite. For all $i \in K_n$ there is some y such that for no z do we have $R(n, (i)_0, p((i)_1), y, z)$. Let y_i be the least such y. $i \to y_i$ is a $\Sigma_2(J_\alpha)$ function (as $p \restriction K_n$ is α-finite) and so is bounded. It follows that from some stage, $f_0(n)[s] \geqslant i(n) \geqslant g(n)$, and that $f_0(n)[s] = i(n)$ for unboundedly many s. Also, $i\colon \omega \to \lambda$ is cofinal.

We now define $f_1(n)[s]$ by induction on s. In general, we let $f_1(n)[s]$ equal $f_0(n)[s]$, unless s is limit and $\liminf_{t \to s} f_1(n)[t] < f_1(n)[s]$; in that case we let $f_1(n)[s] = \liminf_{t \to s} f_1(n)[t]$. Suppose that after s_n, $f_0(n)[s] \geqslant i(n)$; and let C_n be the set of limit points of $\{s > s_n : f_0(n)[s] = i(n)\}$. C_n is a club and $f_1(n)[s] = i(n)$ if $s \in C_n$. Also, $f_1(n)[s] \geqslant i(n)$ for all $s > s_n$. Thus $f_1[s]$ is a weak approximation of i.

Finally, let $f_2(n)[s] = \max\{f_1(k) : k \leqslant n\}$. Then $f_2[s]$ is a weak approximation of $f(n) = \max\{i(k) : k \leqslant n\}$ and is as required. □

We now describe our constructions. The tree of agents is the subtree of $(\alpha \cup \{\texttt{yes}, \texttt{no}\})^{<\lambda}$ of nodes η satisfying the following:

(1) For every even ordinal $i < \operatorname{dom} \eta$, $\eta(i) \in \alpha$, and for every odd ordinal $i < \operatorname{dom} \eta$, $\eta(i) \in \{\texttt{yes}, \texttt{no}\}$.
(2) The set $\{i < \operatorname{dom} \eta : \eta(i) = \texttt{yes}\}$ is finite.

First, all requirements (except for the T_ns) are placed in an effective list of length α. In place 0 we place no requirement.

An agent η of limit length works for no requirement. An agent η of length $\beta + 1$ works for $\eta(\beta)$, if β is even. If β is odd and $\eta(\beta) = \texttt{no}$ then η works for no requirement. If β is odd and $\eta(\beta) = \texttt{yes}$ then η works for T_n, where $n = |\{i < \beta : \eta(i) = \texttt{yes}\}|$. These assignments are fixed throughout the construction, so no initializations are made on account of agents discovering they are working for a new requirement. The restriction on agents exactly implies that for all η, $n(\eta) < \omega$.

The nodes are ordered lexicographically; `yes` is stronger than `no`.

At stage s, we define the path of accessible nodes. Let

$$\delta_0(i)[s] = \begin{cases} p(\beta)[s] & \text{if } i = 2\beta < \lambda, \\ \texttt{yes} & \text{if } i = 2\beta+1 < \lambda \text{ and } \beta \in \operatorname{range} f[s], \\ \texttt{no} & \text{if } i = 2\beta+1 < \lambda \text{ and } \beta \notin \operatorname{range} f[s]. \end{cases}$$

Let $\delta\,[s]$ be the maximal $\eta \subset \delta_0[s]$ such that for all $i < \operatorname{dom}\eta$, $\eta \restriction i$ is a node on the tree of agents.

The true path δ is defined in a similar fashion to $\delta_0[s]$:

$$\delta(i) = \begin{cases} p(\beta) & \text{if } i = 2\beta < \lambda \text{ and } \beta \in \operatorname{dom} p, \\ 0 & \text{if } i = 2\beta < \lambda \text{ and } \beta \notin \operatorname{dom} p, \\ \texttt{yes} & \text{if } i = 2\beta+1 < \lambda \text{ and } \beta \in \operatorname{range} f, \\ \texttt{no} & \text{if } i = 2\beta+1 < \lambda \text{ and } \beta \notin \operatorname{range} f. \end{cases}$$

LEMMA 2.36. *For every $i < \lambda$, $\delta \restriction i$ is a node on the tree of agents. For every $i < \lambda$, $\delta \restriction i \subset \delta\,[s]$ for club many s, and there is some $s_i < \alpha$ after which no $\eta <_\ell \delta \restriction i$ is accessible.*

PROOF. For every $\beta < \lambda$, $p \restriction \beta$ is α-finite; $f \restriction f^{-1}\beta$ is finite and so α-finite; this implies that $\delta \restriction 2\beta$ is α-finite. It mentions only finitely many `yes` answers so it is a node on the tree of agents.

The second and third statements follow from the fact that $p[s] \to p$ is a tame, strong approximation and that $f[s] \to f$ is a tame, weak approximation. Fix $\beta < \lambda$. Let n_β be the least n such that $f(n) > \beta$. Let s_β be a stage such that:

- For all $\gamma \in \beta \cap \operatorname{dom} p$ and $s > s_\beta$, $p(\gamma)[s] = p(\gamma)$.
- For all $k \leqslant n_\beta$ and all $s > s^*$, $f(k)[s] \geqslant f(k)$.
- The set

$$C_\beta = \left\{ s > s_\beta : \begin{array}{c} \forall \gamma \in \beta \setminus \operatorname{dom} p,\ p(\gamma)[s] = 0 \ \& \\ \text{for all } k \leqslant n_\beta,\ f(k)[s] = f(k) \end{array} \right\}$$

  is a (recursive) club.

If $s \in C_\beta$ then $\delta \restriction 2\beta \subset \delta\,[s]$. Also, no $\eta <_\ell \delta \restriction 2\beta$ can be accessible after s_β: say $s > s_\beta$ and $\delta \restriction 2\beta \not\subset \delta\,[s]$; let $\rho = \delta \cap \delta\,[s]$ and let $i = \operatorname{dom}\rho$. If $i = 2\beta$ then since

$$\delta\,[s](i) = p(\beta)[s] \geqslant p(\beta) = \delta(i)$$

we have $\delta <_\ell \delta\,[s]$. Suppose that $i = 2\beta+1$. $\rho \subset \delta, \delta\,[s]$ implies that $\operatorname{range} f \cap \beta = \operatorname{range} f[s] \cap \beta$. If $f(n)[s] = \beta$ then $f(n) \leqslant f(n)[s]$ must also equal β. So we must have $f(n)[s] > \beta = f(n)$ so $\delta(i) = \texttt{yes}$ and $\delta\,[s](i) = \texttt{no}$ and again $\delta <_\ell \delta\,[s]$. □

We want to show that the fairness assumption holds. As usual, we first show that for all η on the true path, $\texttt{init}(\eta)[\alpha] < \alpha$ and that η eventually stops initializing all weaker nodes. Assume $\eta \subset \delta$ and that this holds for all $\rho \subsetneq \eta$.

Say that after s_η, no node to the left of η is ever accessible. The set of agents to the left of η ever accessible is α finite, and each acts at most once after s_η. $\varrho_\alpha = \alpha$ and so the set of such nodes that ever act after s_η is α-finite. It follows that after some stage, η does not get initialized by some node to its left. By the induction assumption, we have $\texttt{init}(\eta)[\alpha] < \alpha$.

To show that after some stage, η does not initialize weaker agents, it is enough, by lemma 2.11, to show that if η works for a D, N or a Z requirement, then η eventually stops initializing on behalf of agents $\rho \subsetneq \eta$. We know that for each $\rho \subsetneq \eta$ there is a stage after which η does not initialize on ρ's behalf; call the least such stage t_ρ. $\rho \to t_\rho$ is $\Pi_1(J_\alpha)$ on $\operatorname{dom}\eta$ and so bounded, as required.

Finally, we need to show that for all $\eta \subset \delta$, $\mathtt{Rest}(\eta)[s]$ is bounded on a recursive club. In fact, we know that for all $\rho \subsetneq \eta$, $r(\rho)[s]$ is either eventually constant or 0 on a recursive club. The set K of $\rho \subsetneq \eta$ such that $r(\rho)[s]$ is eventually constant is $\Sigma_2(J_\alpha)$, hence α-finite. The function taking $\rho \in K$ to that constant value is also $\Sigma_2(J_\alpha)$, hence α-finite (and bounded); so is the function taking $\rho \in K$ to the least stage u_ρ from which $r(\rho)$ is constant.

Let, for $\rho \in K$, $D_\rho = \alpha \setminus u_\rho$, and for other $\rho \subsetneq \eta$, $D_\rho = \{s : r(\rho)[s] = 0\}$. $D = \bigcap_{\rho \subsetneq \eta} D_\rho$ is a club on which $\mathtt{Rest}(\eta)[s]$ is bounded.

CHAPTER 3

Coding Effective Successor Models

In this chapter we combine the construction of effective successor models (see [**NSS98**, Thm. 6.1]) with the tree techniques of section 3.3 and add the construction of an exact pair to code an effective model of arithmetic into the α-r.e. degrees, when α is Σ_2-admissible but $\mathrm{cf}_{\Sigma_3(J_\alpha)}(\alpha) = \omega$. This contrasts with the results of chapter 4.

We first define the notions used. We modify the definition of an SW set to adjust to working above a bottom degree $\mathbf{b}$.

DEFINITION. The *SW set* defined by a quintuple of parameters $\bar{\mathbf{p}} = (\mathbf{b}, \mathbf{p}, \mathbf{q}, \mathbf{r}, \mathbf{l})$ (denoted by $G_{\bar{\mathbf{p}}}$) is the collection of $\mathbf{g} \in (\mathbf{b}, \mathbf{r})$ which are minimal solutions above $\mathbf{b}$ of the inequality $\mathbf{g} \vee \mathbf{p} \geqslant \mathbf{q}$. For $\mathbf{g}_0, \mathbf{g}_1 \in G_{\bar{\mathbf{p}}}$, we let $\mathbf{g}_0 \leqslant_{\bar{\mathbf{p}}} \mathbf{g}_1$ if $\mathbf{g}_0 \leqslant \mathbf{g}_1 \vee \mathbf{l}$.

Next, the definition of the correctness condition $\chi_{\mathtt{SW}}$ and the associated notions which define a model of Robinson arithmetic coded by $\leqslant_{\bar{\mathbf{p}}}$ are taken from the beginning of chapter 2 verbatim, but with $G_{\bar{\mathbf{p}}}$ according to the new definition (working above $\mathbf{b}$).

DEFINITION. Let $\bar{\mathbf{p}}$ be a quintuple satisfying $\chi_{\mathtt{SW}}$. We say that $M_{\bar{\mathbf{p}}}$ is an *effective successor model* if the following two conditions hold.

(1) There is some quadruple $\bar{\mathbf{e}} = (\mathbf{e}_0, \mathbf{e}_1, \mathbf{f}_0, \mathbf{f}_1)$ such that for all $\mathbf{x} \in M_{\bar{\mathbf{p}}}$, if $(\mathbf{x} > 0)^{M_{\bar{\mathbf{p}}}}$, $i < 2$ and $M_{\bar{\mathbf{p}}} \models$ "$\mathbf{x} = i \mod 2$", then

$$\mathbf{x} = (\mathbf{e}_i \vee (\mathbf{x} - 1)^{M_{\bar{\mathbf{p}}}}) \wedge \mathbf{f}_i.$$

(2) For every $\mathbf{x} \in M_{\bar{\mathbf{p}}}$, the set

$$\{\mathbf{y} \in M_{\bar{\mathbf{p}}} : \mathbf{y} <^{M_{\bar{\mathbf{p}}}} \mathbf{x}\}$$

has a least upper bound (in $\mathcal{R}_\alpha$) which we denote by $\sum_{\bar{\mathbf{p}}} \mathbf{x}$. Further, for all $\mathbf{y} \in M_{\bar{\mathbf{p}}}$ such that $\mathbf{y} \geqslant^{M_{\bar{\mathbf{p}}}} \mathbf{x}$ we have $\mathbf{y} \nleqslant \sum_{\bar{\mathbf{p}}} \mathbf{x}$.

We let $\chi_{\mathtt{effective}}(\bar{\mathbf{p}}, \bar{\mathbf{e}})$ state that $\bar{\mathbf{p}}$ satisfies $\chi_{\mathtt{SW}}$ and that $\bar{\mathbf{e}}$ witnesses that $M_{\bar{\mathbf{p}}}$ is an effective successor model. Note that $\chi_{\mathtt{effective}}$ can indeed be stated in a first-order fashion.

Also, we note that if $\bar{\mathbf{p}}$ satisfies $\chi_{\mathtt{SW}}$, $M_{\bar{\mathbf{p}}}$ is standard and the elements of $M_{\bar{\mathbf{p}}}$ form an independent set, then (2) of the definition holds automatically, so (1) is really the interesting requirement for effectiveness of the model, as far as constructions go. The second condition will be used in chapter 4.

We also recall that if $\mathcal{H}$ is an independent set of degrees, then an *exact pair* for $\mathcal{H}$ is a pair of degrees $\mathbf{c}_0, \mathbf{c}_1$ such that the intersection of the principal ideals determined by $\mathbf{c}_0$ and $\mathbf{c}_1$ is exactly the ideal generated by $\mathcal{H}$; namely, for all $\mathbf{x}$,

$$\mathbf{x} \leqslant \mathbf{c}_0 \quad \& \quad \mathbf{x} \leqslant \mathbf{c}_1$$

iff there is some finite $\mathcal{F} \subset \mathcal{H}$ such that

$$\mathbf{x} \leqslant \bigvee \mathcal{F}.$$

In this chapter we prove:

THEOREM 3.1. *Suppose that α is Σ_2-admissible and that $\mathrm{cf}_{\Sigma_3(J_\alpha)}(\alpha) = \omega$. Let $\mathbf{u}$ be a promptly permitting α-r.e. degree. Then there is a quintuple $\bar{\mathbf{p}}$, a pair $\mathbf{c}_0, \mathbf{c}_1$ and a quadruple $\bar{\mathbf{e}}$, all below $\mathbf{u}$, such that $M_{\bar{\mathbf{p}}}$ is a standard effective successor model (witnessed by $\mathbf{e}$), and such that $\mathbf{c}_0, \mathbf{c}_1$ form a minimal pair for $M_{\bar{\mathbf{p}}}$.*

For the definition of prompt permitting and a discussion of how it is used in our construction, see appendix E. We also remark that the theorem can be proven for admissible ordinals α such that $\varrho^2_\alpha = \omega$; the construction is similar, except that the trees of agents that we use are ω (if $\varrho_\alpha = \omega$ or $2^{<\omega}$ (if $\varrho^2_\alpha = \omega < \varrho_\alpha$); assignments of requirements to nodes and the accessible nodes are defined as in sections 3.1 and 3.2.

1. Construction

We are given a promptly permitting set U. We are also given an α-finite partial ordering $\preccurlyeq$ on ω, together with an infinite, α-finite set of $\preccurlyeq$-minimal elements H which we enumerate as $\{h_n : n < \omega\}$. We construct sets $B, P, Q, L, E_0, E_1, C_0, C_1$ and G_n (for $n < \omega$); we require that for all $n < \omega$, $G_n^{[0]} = B$. We denote G_{h_n} by H_n. We let $R = \oplus_{n<\omega} G_n$, $F_0 = \oplus_{n<\omega} H_{2n}$ and $F_1 = \oplus_{n<\omega} H_{2n+1}$.

The requirements are:

(1) T_n: $Q \leqslant_\alpha G_n \oplus P$.
(2) $M_{\Theta,\Phi,W}$: If $\Theta(R) = W$ and $\Phi(B \oplus W \oplus P) = Q$ then there is some j such that $G_j \leqslant_\alpha B \oplus W$.
(3) $D_{j,\Psi}$: $\Psi(\oplus_{i \neq j} G_i) \neq G_j$.
(4) $N_{i,j,\Psi}$: If $j \not\preccurlyeq i$ then $\Psi(G_i \oplus L) \neq G_j$.
(5) $K_{\Xi,K}$: If for all $x \in K$ there are unboundedly many s such that $x \in \Xi(R \oplus P \oplus Q \oplus L)[s]$ then from some s onward, $K \subset \Xi(R \oplus P \oplus Q \oplus L)[s]$ is correct.
(6) $Y_{n,\Phi}$ for $n > 0$: Let $i = n \mod 2$. If $\Phi(H_{n-1} \oplus E_i) = \Phi(F_i)$ is total then $\Phi(F_i) \leqslant_\alpha H_n$.
(7) A_Φ: If $\Phi(C_0) = \Phi(C_1)$ is total then for some n, $\Phi(C_0) \leqslant_\alpha \oplus_{k \leqslant n} H_k$.

The positive ordering requirements (which are met by permitting and coding techniques), are:

- If $j \preccurlyeq i$ then $G_j \leqslant_\alpha G_i \oplus L$.
- If $n > 0$ and $i = n \mod 2$ then $H_n \leqslant_\alpha H_{n-1} \oplus E_i$.
- All sets constructed are recursive in U.

Let $\lambda = \varrho^2_\alpha$. Let $p \colon \lambda \to \alpha$ be partial, 1-1, onto and $\Sigma_2(J_\alpha)$; let $p[s]$ be a recursive, strong, tame approximation to p (see proposition 2.31 and the discussion after claim 2.32). Let $f \colon \omega \to \lambda$ be increasing and cofinal, with a recursive weak approximation $f[s]$, such that each $f[s]$ is increasing (see lemma 2.34).

1.1. The Machine. We use a tree of strategies (a subset of $\alpha^{<\lambda}$) which is almost identical to the tree of section 3.3; nodes of the tree guess the values of p and f. However, we also treat the tree as a pinball machine, where certain nodes act as holes from which balls are released; other nodes act as gates at which balls may reside.

As in section 3.3, only nodes which guess only finitely many elements to be in the range of f are considered as agents on the tree. An agent η which is responsible for guessing if some number ϵ is in range f, and guesses in the affirmative, is called a *pregate*. A *block* consists of the collection of nodes which lie between two pregates. For simplicity, in this construction we let all nodes in each block work on the same kind of requirement; we will need to show that every for requirement R there is indeed a node on the true path working for R. Luckily, elementary arithmetic suffices: all we need to do is to require that for all $n < \omega$, $f(n+1) > f(n) \cdot 2$. This will ensure that the blocks on the true path are long enough so that no requirement is missed. We can easily arrange for that, as λ is closed under addition.

We only assign tasks to agents of successor length. Let η be an agent of length $\beta + 1$.

(1) If β is even then we let η guess the range of f. A node η is a pregate if $\eta(\beta) = 0$.

DEFINITION. An *agent* is a node $\rho \in \alpha^{<\lambda}$ such that there are only finitely many pregates $\eta \subset \rho$. If ρ is an agent then we let $n(\rho)$ be the number of pregates properly contained in ρ.

If η is a pregate then we let η work for $T_{n(\eta)}$.

(2) Suppose that β is odd. Suppose that $(n(\eta))_0 = \langle X, k\rangle$. We have the following cases:
- If $X = M$ and $\eta(\beta) = (\Theta, \Phi, W)$ then η works for $M_{\Theta,\Phi,W}$.
- If $X = K$ and $\eta(\beta) = (\Xi, K)$ then η works for $K_{\Xi,K}$.
- If $X = D$ and $\eta(\beta) = \Psi$ then η works for $D_{k,\Psi}$.
- If $X = N$, $k = \langle i, j\rangle$ and $\eta(\beta) = \Psi$ then η works for $N_{i,j,\Psi}$ (Note that $i, j \leqslant k \leqslant n(\eta)$).
- If $X = Y$ and $\eta(\beta) = \Phi$ then η works for $Y_{k,\Phi}$.
- If $X = A$ and $\eta(\beta) = \Phi$ then η works for A_Φ.

(We use $(n(\eta))_0 = \langle X, k\rangle$ (rather than $n(\eta) = \langle X, k\rangle$) because we want each $\langle X, k\rangle$ to appear infinitely often on the true path.)

For every node η, we let $\epsilon(\eta)$ be the ordertype of nodes preceding η which work for the same kind of requirement as does η. To formally define "same kind":
- All K requirements are of the same kind.
- All M requirements are of the same kind.
- For any j, all $D_{j,\Psi}$ requirements are of the same kind.
- For any i, j, all $N_{i,j,\Psi}$ requirements are of the same kind.
- For any n, all $Y_{n,\Phi}$ requirements are of the same kind.
- All A requirements are of the same kind.

Also, all nodes of odd, non limit length (these are the codes which guess the range of f) are of the same kind.

$\delta[s]$, the *path of accessible nodes*, is calculated by induction; the induction halts at a limit level when we reach a node which is not an agent ($n(\eta) = \omega$). Suppose that $\eta \subset \delta[s]$; we want to find which immediate successor of η is also accessible at s (assuming we did not halt the definition of $\delta\,[s]$); there are two cases. Let $\beta = \operatorname{dom} \eta$ and $\epsilon = \epsilon(\eta^\frown 0)$.

(1) If β is even then $\eta^\frown 0$ is accessible at s if $\epsilon \in \operatorname{range} f\,[s]$; otherwise, $\eta^\frown 1 \subset \delta\,[s]$ (that is, we define $\delta(\beta) = 0\,[s]$ in the first case, 1 in the other).
(2) If β is odd then $\eta^\frown p(\epsilon)\,[s] \subset \delta\,[s]$.

Definition. A *Y-gate* is a pregate ρ such that the nodes in the block which ends with ρ work for a Y requirement (that is, $(n(\rho))_0 = \langle Y, k\rangle$ for some k). An *A-gate* is a pregate such that the nodes in the block which ends with ρ work for an A requirement.

1.1.1. *Restraint.* Let $M = M_{\Theta,\Phi,W}$ be a minimality requirement. A number y is *M-confirmed* at s if

$$\Phi(B \oplus W \oplus P, y)\downarrow = Q(y)\ [s]$$

with use $\zeta \oplus \sigma \oplus \pi = \phi(B \oplus W \oplus P, y)\,[s]$, and

$$\sigma \subset \Theta(R)\,[s].$$

The length of agreement is

$$\ell(M)[s] = \max\{z \mid \forall y < z\ [y \text{ is } M\text{-confirmed at } s]\}.$$

We also let

$$\ell(Y_{n,\Phi})[s] = \max\{\beta : \Phi(H_{n-1} \oplus E_i; \beta)\downarrow = \Phi(F_i; \beta)\downarrow\ [s]\}.$$

Similarly, we let

$$\ell(A_\Phi)[s] = \max\{\beta : \Phi(C_0; \beta)\downarrow = \Phi(C_1; \beta)\downarrow\ [s]\}.$$

Let η be an agent which works for requirement X of type M, Y or A. A stage s is *η-expansionary* if η is accessible at s and for all $t < s$ at which η was accessible, $\ell(X)[t] < \ell(X)[s]$.

Again suppose that η is an agent working for an M, Y or A requirement, and suppose that η is accessible at s. At s, η imposes *restraint* $r(\eta)$ on weaker nodes. The restraint is 0 if s is η-expansionary, or is a limit of η-expansionary stages. Otherwise, we let the restraint be

$$\sup\{t < s : t \text{ is } \eta\text{-expansionary}\}.$$

For agent η, the restraint $\mathtt{Rest}(\eta)\,[s]$ imposed on η at s is

$$\sup_{\eta' \subsetneq \eta} r(\eta')\,[s]$$

(where for an agent η' working for a requirement which is not M, Y or A, $r(\eta') = 0$ always.) We emphasize that in calculating the restraint on η we only consider nodes properly contained in η and not other nodes which are stronger than η (nodes which lie to η's left); these nodes impose restraint by means of initialization).

For shorthand, we let $\chi\,[s] : \operatorname{dom} \delta\,[s] \to \alpha$ be defined by

$$\chi(\beta) = r(\delta \restriction \beta)\,[s].$$

We often say that *the pair η, ξ is accessible at s* if $\eta \subset \delta\,[s]$ and $\xi \subset \chi\,[s]$.

1.1.2. *Followers and Balls.* A node η working for a D or an N requirement attempts to meet the requirement by appointing *followers*. A follower x for a node η working for $D_{j,\Psi}$ is *realized* at s if $\Psi(\oplus_{i\neq j} G_i, x)\downarrow = 0\,[s]$; a follower x for a node η working for $N_{i,j,\Psi}$ is realized at s if $\Psi(G_i \oplus L, x)\downarrow = 0\,[s]$.

As we'll describe shortly, some of these followers will develop long entourages and will trickle down the path of nodes which lie below η in that part of the construction which is the pinball machine. To avoid interfering with infimum requirements (Y, A, M), such followers need to guess the restraint imposed by various nodes preceding η. The classical construction would have the tree do the guess work by adding nodes which guess the outcome $r(\eta')$ of every agent η'. However, in the setting of admissible recursion theory (even under the assumption of Σ_2-admissibility), this approach unfortunately produces problems: it would make the true path too complicated so that it may have length less than λ. We thus need to leave the individual guesswork to every follower. If x is a follower for η, then we let $\xi(x) = \chi\,[s] \upharpoonright \operatorname{dom}\eta$, where s is the stage at which x is appointed. For any other ball z in x's entourage (this notion is described shortly) we let $\xi(z) = \xi(x)$.

To get the reducibilities $H_n \leqslant_\alpha E_i \oplus H_{n-1}$, to every ball (number) which is targeted for H_n, we need to assign a *trace* which is targeted for H_{n-1} or for E_i. The resulting string of balls $y_0, y_1, y_2, \ldots, y_k$ is called an *entourage*; y_0 is the follower and y_{i+1} is y_i's trace. In a sense, the entire entourage, albeit changing in its elements, is the basic unit on the machine. For example, all balls in the same entourage have the same priority (that of the follower), and when a stronger ball or agent receives attention, all balls of the entourage are cancelled (removed from the machine).

There are four possible situations in which an entourage may find itself. Suppose that η is an agent which targets its followers for H_n, $n > 0$.

(1) η has appointed a follower x, which gets a trace t targeted for E_i ($i = n \bmod 2$). The follower and its trace are waiting at η.
(2) x is realized; the entourage is waiting, in a descending fashion, at gates below η. In more detail: the entourage $y_0, \ldots, y_k$ is partitioned into segments $(y_0, \ldots, y_{l_1}), (y_{l_1+1}, \ldots, y_{l_2}), \ldots, (y_{l_{m-1}+1}, \ldots, y_{l_m})$; for $j < m$ we let $\bar{y}_j = (y_{l_j+1}, \ldots, y_{l_{j+1}})$ (where $l_0 = -1$). Each segment $\bar{y}_0, \ldots, \bar{y}_{m-2}$ (which in fact consists of one or two balls) is waiting in a *corral* which lies just behind a gate $\rho \subset \eta$; $\bar{y}_{j+1}$ lies below $\bar{y}_j$. The last segment $\bar{y}_{m-1}$ is located at a gate, at the same height at which $\bar{y}_{m-2}$ is waiting, or below it. The last ball of the entourage is targeted either for some E_j or for H_0.
(3) The entourage is partitioned as before, and as before all but the last segment wait in corrals in a descending fashion. The last segment is lying at the bottom of the machine in the *permitting bin*. The next to last ball in the entourage is targeted for H_0 or for some E_j, and the last ball is targeted for B. At most one ball in this last segment is targeted for some H_n.
(4) The entourage is partitioned as before and again all segments but the last lie in corrals in a descending order. The last segment (which again has a last trace targeted for B) is waiting in some *cage* which lies behind some pregate $\rho \subset \eta$. It is not necessarily the case that this pregate ρ lies below (or at the same level of) the corral in which the one-before-last segment of x's entourage lies.

The possible movement balls in an entourage might go through correspond to these positions.

(1) When η is accessible and x is realized, x and t will be dropped from η. They arrive at the highest gate ρ below η. x is placed in ρ's corral, and t is placed at the gate.
(2) The last segment is located at gate ρ, and $\rho, \xi(x) \restriction \operatorname{dom} \rho$ are accessible. This last segment is released from the gate ρ and falls down to the next gate ρ'; all but the last ball of the segment are placed in the corral belonging to ρ', and the last ball is placed in the permitting bin and is appointed a trace, targeted for B. If ρ is the lowest gate below η, then all the balls which were released are placed in the permitting bin and are assigned a trace targeted for B.
(3) At some stage, the follower x is permitted by U. There are two possibilities:
    (a) There is no ball targeted for some H_n among the balls in x's entourage which are waiting in the permitting bin. In this case all balls in the permitting bin are enumerated into their target sets. The new last ball of the entourage is now waiting in some corral below η; it rolls out to the corresponding gate, and is assigned new traces (as will be described shortly).
    (b) There is one ball y in x's entourage, which is waiting in the permitting bin and is targeted for H_n. y is enumerated into $C_0^{[n]}$, but the ball y is not discarded from the entourage. The last trace (for B) is enumerated into B (and the ball is discarded). The balls left in the bin receive a new trace for B and are all placed in the cage which lies just behind the unique pregate $\rho \subset \eta$ such that $n(\rho) = n$. [The tracing instructions will show that balls in an entourage associated with hole η are only targeted for H_k such that $k < n(\eta)$, so such a pregate exists.]
(4) $\rho, \xi(x) \restriction \operatorname{dom} \rho$ are accessible. If x is not promptly permitted by U then x and the rest of its entourage are cancelled. If it is, then the balls which were waiting in the cage are released and roll down to the bottom of the machine. The unique ball y which was waiting at ρ's cage and which is targeted for some H_n is enumerated into $C_1^{[n]}$ and into H_n; the rest of the balls are enumerated into their target sets. If the follower x was not just enumerated into its target set, then as in case (3)(a), the new last ball of the entourage rolls out from a corral and is assigned new traces. If x was enumerated then η declares victory and cancels all other followers which belong to η.

REMARK 3.2. Every entourage receives attention finitely many times. In particular, the composition of the entourage and the balls' location on the machine are well-defined at limit stages.

We said (cases (3)(a) and (4)) that when a new last ball of the entourage rolls from the corral into the gate, it is assigned new traces. The traces are appointed not to interfere with the preservation that is performed by nodes in the block below the gate. Suppose that ρ is a gate and nodes in the block below ρ are working for $Y_{m,\Phi}$; we do not want balls targeted for both $H_{m-1} \oplus E_j$ and F_j ($j = m \mod 2$) to lie at the gate at the same time (unless they are targeted for the intended infimum H_m). This can always be done. Suppose that a ball x, targeted for H_n, just rolled out to the gate. Let $i = n \mod 2$.

- If $m = n$ then ρ does not care about x; a trace targeted for E_i is fine.
- If $m \neq n$ but $i = j$, then x is targeted for F_i; we cannot appoint a trace for E_i. But we can appoint a trace y_0, targeted for H_{n-1} (which is different from H_{m-1}) and appoint a trace y_1 for y_0, targeted for E_{1-i}.
- If $m \neq n$ and $i \neq j$ then a trace may be targeted for $E_{1-j} = E_i$, about which ρ does not care.

If ρ is an A-gate then ρ's only wish is that two balls targeted for H_ns will not cross at the same time. Thus when a ball waiting at ρ's corral rolls out to the gate, it is assigned a trace for E_i.

To achieve the technical condition which prohibits more than one ball targeted for some H_n to lie in an entourage segment which is waiting at the permitting bin, we stipulate that the lowest gate is always an A-gate.

We remark that followers which are targeted for G_n, $n \notin H$, do not need any traces, and so, when released from their hole η, they fall immediately to the permitting bin, where they do not receive a trace for B. These balls are considered as all other balls as far as permission from U is concerned; when permitted (parallel to case (3)(a) above) they are enumerated into their target set and the agent η declares victory.

We also remark that balls targeted for H_0 do not need traces; nevertheless, whether followers or traces, they need to participate in the machine. For example, when a follower for H_0 is realized, it is only dropped to the highest gate below its hole; only when its guess ξ is deemed correct is it allowed to roll down to the bin, where it needs a trace for B. When permitted it is enumerated into C_0, gets a new trace for B, and is placed in the lowest pregate, awaiting prompt permission.

The positive ordering requirement $G_j \leqslant_\alpha G_i \oplus L$ (if $j \preccurlyeq i$) is met by applying the classical strategy. If a follower x for a $D_{j,\Psi}$ requirement is enumerated into G_j, then it is also enumerated into L. If a follower x for an $N_{i,j,\Psi}$ requirement is enumerated into G_j then it is also enumerated into every G_k for k such that $j \preccurlyeq k$. As is in the classical construction, this does not disturb the machine part because the h_ns are all $\preccurlyeq$-minimal.

Whenever any trace y (which is not a follower) is enumerated into a set H_n then it is also enumerated into L. Again as is in the classical case, this does not disturb the realization of y's follower even if the associated requirement is an N-requirement; this is because y is appointed after x is realized (the first trace appointed for x is targeted for E_i and not for any H_n) and so is too large to disturb that realization.

1.1.3. *Priority.* A follower x is *permitted* at s if some $y \leqslant x$ enters U at s; it is *promptly permitted* at s if some $y \leqslant x$ enters U by stage $p_U(s)$, where p_U is a recursive function witnessing the fact that U permits promptly, with respect to the various r.e. sets we use in the verifications. See the remarks after the proof of lemma E.2.

Just as in the classical construction, for any hole η, we can calculate a finite number which is the number of permissions (either regular or prompt) a follower issued by η needs before it is enumerated into its target. We can therefore assign (shifting) priorities as is done in [**NSS98**]. The priority of x is determined first by the lexicographic priority of η, then by the lexicographic priority of $\xi(x)$, next by the number of permissions it has already received (the more the merrier), and finally by its size (which is the same as by the date of birth, as new followers are chosen large). As in the classical construction, when a follower receives permission it might need to cancel other followers for the same hole, which were previously stronger.

As usual, if some node η' which *lies to the left* of η (i.e. is smaller with regard to the lexicographic ordering on the tree induced by the natural ordering on α) is accessible, or balls issued by η' receive attention, then η is initialized and all of its followers are cancelled (removed from the machine and never considered again). Also as usual, if a ball x issued by η receives attention then all balls issued by η which are weaker than x (according to the priority system just described) are cancelled.

However, we also cancel a ball x which belongs to node η at a stage s at which the following occurs: at s, there is a node $\eta' \subsetneq \eta$ which is accessible, $r(\eta')\,[s] = 0$ but $\xi(x)(\operatorname{dom}\eta') > 0$. It makes sense to do so, because at s we discover that x's guess about the final value of $r(\eta')$ (this is the liminf of $r(\eta')[s]$) is incorrect: either $r(\eta') = 0$ is the correct guess; but even if not, then $r(\eta') \geqslant s$ and clearly $\xi(x)(\operatorname{dom}\eta') < s$. When cancelling a follower x for this reason, we do not initialize η or any weaker agents, nor do we cancel other followers for η which may be weaker than x but whose guess is not yet found to be incorrect.

The following fact, which is true in the classical construction, has even greater importance here.

LEMMA 3.3. *Let y, z be two balls which are on the machine at some stage s and which belong to different entourages. Then y is stronger than z (at s) iff $y < z$.*

PROOF. Say that $y < z$. Since new balls are chosen large, y is appointed before z. When z is appointed, y is stronger than z (or y gets cancelled immediately). z in fact remains weaker than y as long as y is left on the machine, since z's strengthening beyond y will immediately lead to y's cancellation. Note also that as long as z is on the machine, y cannot receive attention. □

This shows that at any given stage, the priority ordering on balls is a well-ordering. During the construction we will need to pick, among a perhaps infinite set of balls which require attention, the strongest such ball, and this gives us the justification to do so.

1.1.4. *SW Ingredients.* As in the classical construction, the part of the construction which is devoted to building the SW set parameters (namely those agents which work for K, M and T requirements) has very little interaction with the machine. We can therefore make the same definitions for these elements of the construction, remembering that we are working above B.

T Requirements. Suppose that pregate ρ is assigned to T_n. ρ defines a functional Γ_ρ, with the intention of having $\Gamma_\rho(G_n \oplus P) = Q$. ρ may only add an axiom to Γ_ρ when it is accessible.

All uses of Γ functionals are successor ordinals; $\gamma_\rho(x) = \gamma_\rho(G_n \oplus P, x)$ denotes the length of the use of the computation $\Gamma_\rho(G_n \oplus P, x)$ minus one, so enumerating it into G_n or P destroys the computation.

Chits. Suppose that agent η works for $M = M_{\Theta,\Phi,W}$. η defines weak functionals $\Delta_{\eta,j}$ for every $j \leqslant n(\eta)$, with intended oracle $B \oplus W$; the intention is that if the hypothesis of M holds then for some $j \leqslant n(\eta)$ we'll have $\Delta_{\eta,j}(B \oplus W) =^* G_j$.

η is allowed to extend the definition of $\Delta_{\eta,j}(B \oplus W)$ (i.e. to enumerate a new axiom into $\Delta_{\eta,j}$) at stages which are η-expansionary. As is implies by the intention, η always defines the value of $\Delta_{\eta,j}(B \oplus W, x)[s]$ to be $G_j(x)[s]$.

η makes $\Delta_{\eta,j}$ monotone, so that at any stage we'll have $\operatorname{dom} \Delta_{\eta,j}(B \oplus W)$ an ordinal. If η wishes to extend $\Delta_{\eta,j}$ at s then it defines one new axiom on the input $x = \operatorname{dom} \Delta_{\eta,j}(B \oplus W)\,[s]$ (we'll ensure that the use is larger than the uses $\delta_{\eta,j}(B \oplus W, y)\,[s]$ for all $y < x$ so that indeed monotonicity is maintained.)

To each computation $(\zeta \oplus \sigma; x, l) \in \Delta_{\eta,j}$ is associated a *chit* (y, π) (we sometimes also refer to y as the chit). When η wishes to define $\Delta_{\eta,0}(x)$ at stage s, it picks a new *suitable* chit. A chit (y, π) is suitable to be picked for a new computation $\Delta_{\eta,0}(B \oplus W, x)\,[s]$ if:

(1) $y \in \alpha^{[\eta]}$.
(2) $y < \ell(M)[s]$.
(3) $\phi(B \oplus W \oplus P, y)[s] = \zeta \oplus \sigma \oplus \pi$.
(4) $y >$ `init`$(\eta)[s]$ and $y > t$ for any $t < s$ at which η defined any $\Delta_{\eta,0}$ computation.

If s is η-expansionary and there is a suitable chit y, then it defines $\Delta_{\eta,0}(B \oplus W, x)$ with use $\zeta \oplus \sigma$.

Now let $j > 0$. Suppose that at s, η wishes to define $\Delta_{\eta,j}(B \oplus W, x)$. To do so, it needs to find a chit (y, π) which is suitable to be picked for this computation (we also say that the chit is *j-suitable*). The suitability conditions are:

(1) (y, π) is a chit for a computation $\Delta_{\eta,j-1}(B \oplus W, x')[s]$ (whose use is $\zeta \oplus \sigma$).
(2) Further, that chit is still *active*, which means that $\pi \subset P\,[s]$ (an inactive chit is also called *cancelled*).
(3) The computation $\Delta_{\eta,j-1}(B \oplus W, x')[s]$ is *failed*, which means that its value disagrees with $G_{j-1}(x')$ (necessarily x' entered G_{j-1} at some stage after the computation was defined).
(4) The size condition: $y > t$ for any $t < s$ at which η defined any $\Delta_{\eta,j}$ computation.

If such a chit is found then η defines $\Delta_{\eta,j}(B \oplus W, x)[s]$ with the same use $\zeta \oplus \sigma$ as the accompanying computation $\Delta_{\eta,j-1}(B \oplus W, x')[s]$.

If a computation $\Delta_{\eta,j}(B \oplus W, x)$ which is defined at s becomes incorrect at a later stage t (that is some number smaller than the use enters B or W) then the accompanying chit (y, π) is cancelled and never considered again. Note again that if (y, π) is a chit for $\Delta_{\eta,j}(B \oplus W, x_j)$ for $j \leqslant i$ (where $i \leqslant n(\eta)$) then the uses of all of these computations are the same and so all such computations disappear together). However, (2) shows that the chit may be cancelled even if the computations still hold.

At stage s, η may wish to use a chit (y, π) for purposes of victory. Suppose that $\rho \subset \eta$ is a pregate which works for T_n ($n(\rho) = n$). A chit (y, π) is *cleared* by ρ at stage s if $\neg(\Gamma_\rho(B \oplus G_n \oplus P, y) \downarrow= 0[s])$, or if $\Gamma_\rho(B \oplus G_n \oplus P, y) \downarrow= 0[s]$ with use $\gamma_\rho(y)[s] > \operatorname{dom} \pi$. The chit y is *victorious* if it is M-confirmed at s, still active, is greater than `Rest`$(\eta)\,[s]$, and is cleared by all pregates ρ below η.

Pointers. An agent η working for a requirement $K_{\Xi,K}$ keeps a pointer $i(\eta)[s]$, which is the next element of K of which it needs to take care. Unless initialized, at a limit stage s we have $i(\eta)[s] = \min(K \setminus \sup\{i(\eta)[t]\}_{t<s})$. When η is initialized we set $i(\eta) = \min K$. If η acts at s then we update the pointer: $i(\eta)[s+1] = \min(K \setminus i(\eta)[s]+1)$. If $i(\eta)[s] = \max K$ and η acts at s then η declares victory.

1.2. Construction. Each stage of the construction is divided into three phases. At the first phase, agents working for lowness requirements act; as in the construction of chapter 2, this is done even before $\delta\,[s]$ is calculated because of the incompatibility between lowness and the tree construction. At the second phase, the path of accessible nodes is computed and then finitary-type action takes place: balls moving on the machine, numbers enumerated into sets, new followers being appointed, agents declaring victory. At the third phase, agents along $\delta\,[s]$ enumerate new axioms into the functionals they are building (but further initializations may occur).

We emphasize that at any stage s (and any phase of that stage), if η is an agent which declared victory at some stage $t < s$ and η was not initialized between t and s, then η does not require attention at s and does not make any action.

As in chapter 2 we let $\texttt{init}(\eta)\,[s]$ be the supremum of stages $t < s$ at which η is initialized.

1.2.1. *First Phase.* An agent η is *allowed to act* at the first phase of stage s if there was a stage $t < s$ at which η was accessible, such that between t and s, neither was η initialized, nor did it act. An agent η on the tree, working for $K_{\Xi,K}$, *wishes to act* at the first phase of s if

$$i(\eta) \in \Xi(R \oplus P \oplus Q \oplus L)\,[s].$$

We let every agent η which is allowed to act and wishes to act initialize all agents on the tree which are weaker than η (again we note that the priority (lexicographic) ordering on agents is linear but not well-founded, so there may not be a strongest node which acts; nevertheless all such η must either act or be initialized, so we let them all act, and of course if there is no strongest one they all get intialized.) We also update $i(\eta)$ as described in section 1.1.4. Also, η may declare victory.

1.2.2. *Second Phase.* First, we calculate the path of accessible nodes $\delta\,[s]$. Agents which are weaker than $\delta\,[s]$ are initialized. We also calculate the restraint which accessible agents impose, and cancel balls whose guess about the restraint imposed is found to be incorrect.

We say that a follower x on the machine (which is not neccessarily issued by an accessible agent; it could be an agent to the left of $\delta\,[s]$) *requires attention* at s if one of the following holds.

(1) The end of x's entourage is waiting at a gate ρ, and $\rho, \xi(x) \restriction \operatorname{dom}\rho$ are accessible.
(2) x is waiting at its appointing hole η and is realized; η is accessible [Note that we do *not* require that $\xi(x)$ be accessible].
(3) The end of x's entourage is waiting in a cage associated with a pregate ρ, and $\rho, \xi(x) \restriction \operatorname{dom}\rho$ are accessible.
(4) The end of x's entourage is waiting in the permitting bin and x is permitted by U at s.

If a follower x which was issued by agent η requires attention then we also say that η requires attention. In addition, an accessible hole η which works for an N or D requirement requires attention if:

(5) No follower for η is waiting at the hole η, and no other follower for η requires attention.

An accessible agent η which works for $M = M_{\Theta,\Phi,W}$ requires attention if:

(6) There is some $\sigma \subset \Theta(R)[s]$ and some $x < \operatorname{dom} \sigma$ s.t. $\sigma(x) = 0$ & $x \in W[s]$, or if

(7) $r(\eta)\,[s] = 0$ and there is a victorious chit y for η which is promptly permitted by U.

We already noticed that the balls on the machine at this phase are well-ordered, and so there is a strongest ball which requires attention. Also, $\delta\,[s]$ is well-ordered, and only accessible nodes require attention independently (i.e. not via any follower). Thus there is, overall, a strongest agent η which requires attention. If some follower for η requires attention then we let x be the strongest such follower. We initialize agents weaker than η and cancel balls weaker than x (note that x may have just received new permission so it may cancel balls which were up to now stronger).

In any case, we let η act according to the relevant case.

(1-4) η acts according to the instructions above (section 1.1.4). Recall that in case 2, if x is not targeted for any H_n then we simply put it in the permitting bin.

(5) η appoints a new, large follower x, and if necessary, appoints a trace.

(6) η initializes weaker requirements and declares "easy" victory.

(7) y is enumerated into Q, and for all pregates $\rho \subset \eta$, if $\Gamma_\rho(G_{n(\rho)} \oplus P, y)\downarrow = 0\,[s]$, then ρ puts $\gamma_\rho(y)\,[s]$ into P. η declares victory.

1.2.3. *Third Phase.* At the third phase, every accessible agent (which was not just initialized) working for M or T may extend the functionals it defines.

An accessible pregate ρ extends Γ_ρ by defining it on the least x such that $\Gamma_\rho(G_{n_\rho} \oplus P, x)\uparrow\ [s]$, with new large use and value $Q(x)\,[s]$.

An accessible agent η which works for $M_{\Theta,\Phi,W}$ is allowed to define new $\Delta_{\eta,j}$ computations only if s is η-expansionary. The definitions are as described above. Further, if η defines $\Delta_{\eta,j}(x)$ and $\eta' \supset \eta$ is an agent which has x as a ball in one of its entourages, then η' initializes all agents weaker than itself.

That's the end of the construction.

2. Verifications

2.1. True Path. The *true path* is obtained effortlessly as in section 3.3, because of the properties of the approximations $p\,[s]$ and $f\,[s]$. However, unlike that section, we do need an inductive definition of δ.

LEMMA 3.4. *Let η be an agent on the tree. Suppose after some stage s^*, no agent to the left of η is ever accessible. Then*

$$\{t \geqslant s^* \,:\, \eta \subset \delta\,[t]\}$$

is closed.

PROOF. Suppose not. Let $t > s^*$ be a stage which is a limit of stages at which η is accessible, and suppose that η is not accessible at t. It is impossible that $\delta\,[t]$ is properly contained in η, because the construction of $\delta\,[t]$ is halted only when we get to a node which is not an agent; η (and every $\eta' \subset \eta$) is an agent. It follows that η lies to the left of $\delta\,[t]$. Let β be the least such that $\eta(\beta) \neq \delta(\beta)\,[t]$; so $\eta(\beta) < \delta(\beta)\,[t]$. Let

$$\epsilon \;=\; \epsilon(\eta \restriction \beta + 1) = \epsilon(\delta \restriction \beta + 1\,[t]).$$

We, of course, have two cases.

(1) β is odd. In this case, we know that the set of stages $s < t$ at which $p(\epsilon)\,[s] = \eta(\beta)$ is unbounded in t; this happens at each stage at which η is accessible, by the definition of $\delta\,[s]$. However, $p(\epsilon)[t] = \delta(\beta)[t] > \eta(\beta)$. This can only happen if $\eta(\beta) = 0$ but the set of s such that $p(\epsilon)[s] = 0$ is closed; contradiction.

(2) β is even. Let $n = n(\eta \restriction \beta)$. $\eta \restriction \beta \subset \delta\,[t]$ implies that η's and $\delta\,[t]$'s opinion on $f \restriction n$ are the same: for all $m < n$, $f(m)[t] = f(m)[s]$ where s is any stage at which η is accessible. However, $\eta(\beta) < \delta(\beta)[t]$ again implies that $\eta(\beta) = 0$, that is, $f(n)[s] = \epsilon$ whenever η is accessible but $f(n)[t] > \epsilon$. But again, if for unboundedly in t many s we have $f(n)[t] = \epsilon$ then we must have $f(n)[t] \leqslant \epsilon$; contradiction. □

LEMMA 3.5. *There is a true path $\delta\colon \lambda \to \alpha$ such that for all $\beta < \lambda$,*

(1) *$\delta \restriction \beta$ is α-finite and is in fact an agent on the tree.*
(2) *$\delta \restriction \beta$ is accessible unboundedly often.*
(3) *There is some stage $s < \alpha$ after which no agent η which lies to the left of $\delta \restriction \beta$ is ever accessible.*

PROOF. By induction on $\beta < \lambda$ we define δ up to β and show that (1-3) hold for the nodes defined.

We first notice that if for some γ, $\delta \restriction \gamma$ is defined and satisfies (1-3), then by lemma 3.4, the set of stages after s (which is given by (3)) at which $\delta \restriction \gamma$ is accessible is a (recursive) club.

Suppose that $\delta \restriction \beta$ is defined and that (1-3) hold for $\delta \restriction \beta$; we define $\delta(\beta)$. Let $\epsilon = \epsilon((\delta \restriction \beta)^\frown 0)$. If β is even, then if $\epsilon \in \operatorname{range} f$ then $\delta(\beta) = 0$, otherwise $\delta(\beta) = 1$. If β is off and $\epsilon \notin \operatorname{dom} p$ then $\delta(\beta) = 0$, otherwise $\delta(\beta) = p(\epsilon)$.

We now show that (1-3) hold for $\delta \restriction (\beta + 1)$. (1) is immediate. For (2), let C be a recursive club of stages at which $\delta \restriction \beta$ is accessible. If β is even then let E be a recursive club of stages at which $\epsilon \in \operatorname{range} f\,[s]$ is correct; if β is odd let E be a recursive club of stages at which $p(\epsilon)[s] = p(\epsilon)$ (or $p(\epsilon)[s] = 0$ if $\epsilon \notin \operatorname{dom} p$). Then on $C \cap E$, $\delta \restriction (\beta + 1)$ is accessible.

For (3), suppose that after s_0, no agent to the left of $\delta \restriction \beta$ is ever accessible. If $\delta(\beta) = 0$ then clearly after s_0 no node to the left of $\delta \restriction (\beta + 1)$ is ever accessible. Otherwise, if β is even, choose n minimal such that $f(n) \geqslant \epsilon$; then $f(n) > \epsilon$ and we can find some $s_1 > s_0$ after which we always have $f(n)[s] > \epsilon$. Then if $s > s_1$ and $\delta \restriction \beta$ is accessible at s, then at s, ϵ is not guessed to be in the ranges of f and so $(\delta \restriction \beta)^\frown 0$ is not accessible at s. As $s_1 > s_0$ it follows that no node to the left of $\delta \restriction (\beta + 1)$ is ever accessible after s_1. If β is odd, then $\epsilon \in \operatorname{dom} p$ and we can find some $s_1 > s_0$ after which we always have $p(\epsilon)[s] = p(\epsilon)$; after s_1, no node to the left of $\delta \restriction (\beta + 1)$ is ever accessible.

Now suppose that β is a limit ordinal and that δ is defined up to β; and that (1-3) hold for $\delta \upharpoonright \gamma$ for all $\gamma < \beta$. We need to verify that (1-3) hold for $\delta \upharpoonright \beta$.

For (1), we show that the function $\gamma \to \delta \upharpoonright \gamma$, defined on β, is recursive, hence α-finite. We can define this function by induction, using the following sets and functions:

- $\beta \setminus \operatorname{dom} p$;
- $p \upharpoonright (\beta \cap \operatorname{dom} p)$;
- $\beta \cap \operatorname{range} f$.

The first is $\Pi_2(J_\alpha)$ and so α-finite (as $\beta < \lambda$); the second is now $\Delta_2(J_\alpha)$ and so α-finite (using Σ_2-admissibility); the third is finite and so α-finite. Now the induction is straightforward; given $\delta \upharpoonright \gamma$, $\epsilon((\delta \upharpoonright \gamma)^\frown 0)$ is computed and $\delta(\gamma)$ is computed as it was defined above we note that always $\epsilon((\delta \upharpoonright \gamma)^\frown 0) < \beta$.

For $\gamma < \beta$ let s_γ be the least stage witnessing (3) for $\delta \upharpoonright \gamma$. As $\gamma \to \delta \upharpoonright \gamma$ is recursive, $\gamma \to s_\gamma$ is $\Pi_1(J_\alpha)$ and so α-finite. For $\gamma < \beta$ let

$$E_\gamma = \{s > s_\gamma : \delta \upharpoonright \gamma \subset \delta[s]\};$$

E_γ is a club, and these clubs are uniformly recursive. It follows that $E = \bigcap_{\gamma<\beta} E_\gamma$ is a club of stages at which $\delta \upharpoonright \beta$ is accessible, so (2) holds. Also, $s_\beta = \sup_{\gamma<\beta} s_\gamma$ witnesses that (3) holds for $\delta \upharpoonright \beta$. □

LEMMA 3.6. *For every requirement R there is some $\beta < \lambda$ such that $\delta \upharpoonright \beta$ is assigned to work for R.*

PROOF. This amounts to showing that for every *kind* of requirement X, the ordertype of the set of $\beta < \lambda$ such that $\delta \upharpoonright \beta$ is assigned to a requirement of kind X, is λ. This is why we required $f(n+1) > f(n) \cdot 2$. For kind X and every n, there is a block of nodes of δ which has length at least $f(n)$ and is devoted to X; so there are at least $f(n)$ many nodes on the true path working for X requirements. As f is cofinal in λ we're done. □

Define $\chi\colon \lambda \to \alpha$ by letting $\chi(\beta) = \liminf_s r(\delta \upharpoonright \beta)[s]$. In fact we know that either there are unboundedly many $\delta \upharpoonright \beta$-expansionary stages, in which case $r(\delta \upharpoonright \beta)[s] = 0$ on a club of stages and $\chi(\beta) = 0$, or from some stage onward, $r(\delta \upharpoonright \beta)[s]$ is constant (is the supremum of all $\delta \upharpoonright \beta$-expansionary stages) and equals $\chi(\beta)$.

LEMMA 3.7. *For all $\beta < \lambda$ the following holds.*

(1) *$\chi \upharpoonright \beta$ is α-finite.*
(2) *$\delta \upharpoonright \beta, \chi \upharpoonright \beta$ are accessible on club many stages.*
(3) *There is some stage t after which for all $\gamma < \beta$, whenever $\delta \upharpoonright \gamma$ is accessible, we have $r(\delta \upharpoonright \gamma)[s] \geqslant \chi(\gamma)$.*

Note that the third condition is stronger than saying that after some stage, whenever $\delta \upharpoonright \gamma$ is accessible, $\chi[s]$ is not lexicographically to the left of $\chi \upharpoonright \gamma$, but is immediately implied by the statement that $\chi``\beta$ is bounded below α (any bound witnesses (3)).

PROOF. Fix $\beta < \lambda$. Let K be the set of $\gamma < \beta$ such that there are unboundedly many $\eta \upharpoonright \gamma$-expansionary stages. K is a $\Pi_2(J_\alpha)$-definable subset of β which in turn is smaller than $\lambda = \varrho_\alpha^2$; it follows that K is α-finite.

Now $\chi \upharpoonright \beta$ is defined as follows:

- If $\gamma \in K$ then $\chi(\gamma) = 0$.
- If $\gamma \notin K$ then $\chi(\gamma) = t$ where t is $\delta \upharpoonright \gamma$-expansionary, or a limit of $\delta \upharpoonright \gamma$-expansionary stages, but there is no $s > t$ which is $\delta \upharpoonright \gamma$-expansionary.

As $\gamma \to \delta \upharpoonright \gamma$ is recursive, we see that $\chi \upharpoonright \beta$ is $\Pi_1(J_\alpha)$ and so α-finite.

Again let s_β be the least stage after which no node to the left of $\delta \upharpoonright \beta$ is ever accessible, and let E_β be the club of stages after s_β at which $\delta \upharpoonright \beta$ is accessible. Let $t_\beta = \sup \chi``\beta$. For $\gamma \in K$, let D_γ be the set of stages after s_β at which $r(\delta \upharpoonright \gamma)[s] = 0$ (i.e. the $\delta \upharpoonright \gamma$-expansionary stages or limits of those). D_γ is a club and the D_γs are uniformly recursive. It follows that

$$E = [t_\beta, \alpha) \cap E_\beta \cap (\bigcap_{\gamma \in K} D_\gamma)$$

is a club of stages at which $\delta \upharpoonright \beta, \chi \upharpoonright \beta$ are accessible. Also, t_β witnesses (3) for β. □

2.2. Fairness.

DEFINITION. An agent $\eta \subset \delta$ on the tree is *treated fairly* if $\texttt{init}(\eta)[\alpha] < \alpha$. If $\beta < \lambda$ and $\delta \upharpoonright \beta$ is treated fairly then we let r^*_β be the maximum of $\texttt{init}(\delta \upharpoonright \beta)[\alpha]$ and the least stage t witnessing (3) of lemma 3.7

Suppose that $\eta = \delta \upharpoonright \beta$ is treated fairly, let $r^* = r^*_\beta$, and just for this subsection let $\chi = \chi \upharpoonright \beta$. We examine how η affects weaker agents. The only agents which initialize weaker agents are agents working for M, K, D and N requirements.

If η works for an M requirement then M initializes weaker agents at most once after r^*.

LEMMA 3.8. *Suppose that η works for $K_{\Xi,K}$. Then after some stage, η does not initialize weaker agents, and the requirement is met.*

PROOF. After $\texttt{init}(\eta)[\alpha]$, $i(\eta)[s]$ is non-decreasing. Let

$$K' = \{i(\eta)[s] \ : \ s > \texttt{init}(\eta)[\alpha]\};$$

K' is α-finite as it is an initial segment of K. The function which takes $x \in K'$ to the least $s > r^*$ at which $i(\eta)[s] = x$ is recursive and so bounded by some s^*. After s^*, η does not initialize weaker agents.

If η acts at $s > \texttt{init}(\eta)[\alpha]$ then $i(\eta)[s] \in \Xi(R \oplus P \oplus Q \oplus L)$ is permanently correct. Thus if η declares victory after $\texttt{init}(\eta)[\alpha]$ then the requirement is positively met.

If not, let i be the final value of $i(\eta)[s]$, stabilizing after s^*. For no $s > s^*$ do we have $i \in \Xi(R \oplus P \oplus Q \oplus L)\,[s]$ and so the requirement is met vacuously. □

Suppose then that η works for a D or an N requirement.

OBSERVATION 3.9. A *non-deficiency* stage for the construction is a stage s such that no ball on the machine at the end of s will ever receive attention again. There are unboundedly many such stages. For if t is any stage, we can let x be the smallest follower ever to receive attention after t and let s be the last stage at which x receives attention. Now after x receives attention at s, all balls left on the machine are at least as strong as x, hence not larger; they will never receive attention again. It may be the case that some of these balls will later be cancelled. If s is a non-deficiency stage for the construction, then all sets constructed are correct up to s at s, and all computations which are left at the end of s will never be destroyed; this is because these computations have use at most s, and the only balls to receive attention after s are those appointed after s, which are larger than s. Note that this applies only to sets constructed by the machine, namely, G_n, C_i, E_i and L; P and Q are excluded from this discussion.

LEMMA 3.10. *Eventually, η stops appointing new followers.*

It follows that there is a stage after which no followers for η receive attention.

PROOF. Assume the contrary; we compute U. We know that no follower is enumerated after r^*. Of course it cannot be the case that some follower x is appointed after r^*, is never cancelled, and is never realized at a stage at which η is accessible. Also, for every follower x appointed after r^*, for all $\beta < \operatorname{dom} \eta$, $\xi(x)(\beta) \geqslant \chi(\beta)$. In fact, if $\xi(x) \neq \chi$ then if not cancelled earlier, x will be cancelled at a stage at which η, χ are accessible. If $\xi(x) = \chi$ then at every stage at which η, χ are accessible and at which x has ball waiting at a gate or in a cage, x will require attention. It follows that no follower can have balls permanently stuck at some gate or in some cage. Thus every follower appointed after r^* is either eventually cancelled, or becomes realized and some balls of its entourage get permanently stuck in the permitting bin.

For shorthand, call a follower x for η such that $\xi(x) = \chi$ a *follower for η, χ*.

We claim that if η appoints unboundedly many followers, then it appoints unboundedly many followers for η, χ. Suppose not; suppose that after stage $s_0 > r^*$, η does not appoint any followers x with $\xi(x) = \chi$. Let $s_1 > s_0$ be a non-deficiency stage. Let $s_2 > s_1$ be a stage at which η appoints some follower x (so at s_2, no follower occupies the hole), and let $s_3 \geqslant s_2$ be any stage at which η, χ are accessible. For all y appointed by η after s_1 we have, by assumption, $\xi(y)$ weaker than χ; it follows that at s_3, all such balls are cancelled. However, all balls appointed before or at s_1 do not require attention after s_1. Thus at s_3, no follower for η requires attention and the hole at η is empty. At s_3 requires attention and receives it (as $s_3 > \texttt{init}(\eta)[\alpha]$) and appoints a new follower – contradiction.

Let $m < \omega$ be the largest number such that unboundedly many followers x for η, χ are permitted m many times. Say that all followers for η, χ appointed after $s_0 > r^*$ are permitted at most m times; letting $s_1 > s_0$ be a non-deficiency stage for the construction, we know that all followers x for η, χ which receive attention after s_1 are permitted at most m times. Also, we note that after s_1, only followers appointed after r^* receive attention, and so if x is a follower which receives attention after s_1, then for all $\beta < \operatorname{dom} \eta$, $\xi(x)(\beta) \geqslant \chi(\beta)$.

CLAIM 3.11. There are unboundedly many followers x for η, χ which at some $s > s_1$ are permitted for the mth time and are never later cancelled *by other balls.*

Note that followers for η, χ have the correct guess so they can only be cancelled by stronger balls, or by asking for prompt permission and being denied that permission.

PROOF. For any non-deficiency stage $s_2 > s_1$, consider the smallest x which is a follower for η, χ and which receives permission for the mth time at some stage after s_2; suppose this happens at $s_3 > s_2$.

Every ball appointed after s_3 is always going to be weaker than x, as it can never receive more permissions than x. Suppose that at the end of s_3, y is a ball on the machine. At s_3, y is stronger than x and so smaller than x. If y does not belong to η then it belongs to a stronger node; y will not receive attention later since we are after $\texttt{init}(\eta)[\alpha]$. If y is a follower for η then $\xi(y)$ is at least as strong as χ (or y is cancelled at s_3). If $\xi(y)$ is stronger than χ then as mentioned above, y was appointed before r^* and so does not receive attention after s_1. Assume then that y is a follower for η and that $\xi(y) = \chi$.

x not cancelling y at s_3 implies that by the beginning of s_3, y has already received at least m permissions. By minimality of x, the mth permission for y must have been granted before or at s_2. As s_2 is a non-deficiency stage, we know that y does not receive attention after s_2 and so will never cancel x. □

We note that all entourages for η follow exactly the same path down the tree; this is because we let balls wait at gates even if those gates are occupied by stronger balls. It follows that the status of the entourage (last segment waiting at a gate or in a cage) after receiving m permissions is the same for all followers for η. We claim that after receiving m permissions, it cannot be the case that the last segment lies in a cage.

Suppose not; then after receiving the mth permission, followers have balls waiting in the cage behind pregate ρ. There are unboundedly many followers x for η, χ with the following property: At some stage $s_2 > s_1$, x is permitted for the mth time and at the next stage $s_3 \geqslant s_2$ at which $\rho, \chi \upharpoonright \operatorname{dom} \rho$ is accessible, x is still on the machine, requires attention and receives it (any follower which is permitted m times and not cancelled by stronger balls would fit this description). Now none of these balls can be permitted once more, so all are cancelled when at s_3 they ask for prompt permission. However, the set of such xs is recursively enumerable; this contradicts the fact that U is promptly simple.

Let $\varphi(x, s)$ be the statement: "x is a follower for η, χ and $s > s_1$. η, χ are accessible at s. At the end of s, x is on the machine (uncancelled) and has some balls waiting in the permitting bin. Further, before s, x has already been permitted m times." φ is computable.

The analysis above shows that for every follower for η, χ which is permitted for the mth time at some stage after s_1 and not later cancelled by stronger balls, there will be some s such that $\varphi(x, s)$ holds: at the last stage at which such a follower x receives attention, balls of its entourage are placed in the permitting bin where they are to reside for ever. It follows that there are unboundedly many followers x such that at some s, $\varphi(x, s)$ holds.

We show that if $\varphi(x, s)$ holds then x will in fact never be cancelled after s, and so that $U \upharpoonright x\,[s] = U \upharpoonright x$ – this gives us an algorithm to calculate U. Take such x and s. Of course, we know that x will not be permitted again, and so will not receive attention again and will not be cancelled at a later stage due to lack of prompt permission. Also, as noted earlier, x's guess ξ is correct and so x will not be cancelled due to finding its guess is incorrect.

Let $t \leqslant s$ be the stage at which x last receives attention (at t, balls in x's entourage are placed in the bin). Balls which at the end of t are weaker than x are cancelled at t; as above, new balls appointed after t will always be weaker than x since they cannot be permitted more times than x. Consider a ball y which at the end of t is stronger than x. If y is not a follower for η then y cannot receive any attention after t. Suppose that y is a follower for η. We know that if y receives attention after t then $\xi(y)(\beta) \geqslant \chi(\beta)$ for all $\beta < \operatorname{dom}\eta$. If $\chi \neq \xi(y)$ then y is weaker than x and so is cancelled at t. We thus assume that y is a follower for η, χ.

By t, y has already been permitted m times, and so either has balls in the bin (in which case it will never receive attention after t and will thus never cancel x) or has balls waiting at a gate ρ. The latter is impossible though: if that were the case, then at some stage after t, but not later than s, $\rho, \chi \restriction \rho$ is accessible and y requires attention and cancels x. □

COROLLARY 3.12. *η satisfies its requirement.*

PROOF. Suppose that at $s > \mathtt{init}(\eta)[\alpha]$, η enumerates a follower x into its target set and declares victory. At $t < s$ the follower is found to be realized and is dropped from the hole; it cancels weaker balls. Smaller (stronger) balls do not receive attention until s, since that would cancel x. All balls appointed later are larger than t, hence larger than the use of the computation which realizes x. Also, x's first trace is targeted for some E_i and so does not destroy the realizing computation; all other traces are appointed after t so are too large to destroy that computation. Thus x is still realized at s. At s, all other balls associated with η are cancelled and η never acts again. Agents weaker than η are initialized at s and later appoint large balls. Agents stronger than η never act after $\mathtt{init}(\eta)[\alpha]$. It follows that the computation realizing x is preserved for ever and so the requirement is positively met.

Suppose this does not happen. After stage $s_0 > r^*$, η appoints no new followers, and no followers for η receive any attention. There is a stage $s_1 > s_0$ after which no balls associated with η are ever cancelled (the set of such balls is r.e., hence α finite). Let $s_2 > s_1$ be a stage at which η is accessible. At that stage, there must be a follower x for η which occupies the hole, unrealized; otherwise, η would appoint a follower at s_2, as no follower requires attention at s_2. x is never cancelled and never receives attention. It follows that x is not correctly realized – if it is at some stage, then at the next stage at which η is accessible, x requires attention, even if its guess is incorrect (too strong). Thus the requirement is met negatively. □

Suppose that $\rho \subsetneq \eta$ works for $M = M_{\Theta,\Phi,W}$; let $j \leqslant n(\rho)$. What follows is identical to the arguments given in section 2.4.1 (observation 2.10 and its sequel).

OBSERVATION 3.13. Suppose that $s > t > \mathtt{init}(\rho)[\alpha]$. Suppose that $\Delta_{\rho,j}(B \oplus W, x) \downarrow [t]$ with use $\zeta \oplus \sigma$, that $\zeta \subset B[s]$ and that $\sigma \subset \Theta(R)[s]$. Then ρ does not redefine $\Delta_{\rho,j}(x)$ at stage s. For if it does, then $\Delta_{\rho,j}(B \oplus W, x) \uparrow [s]$ which means that $\sigma \not\subset W[s]$ anymore; necessarily some $y < \operatorname{dom}\sigma$ entered W between t and s. But then ρ can get an easy victory at s and defines nothing.

Let $K(\eta)$ be the final set of balls issued by η which are never cancelled (it is α-finite), and let $r^{**} > r^*$ be a stage after which η neither appoints any followers, nor does any follower receive attention or get cancelled. Suppose that $x \in K(\eta)$ and suppose that at some stage $s > r^{**}$, ρ defines $\Delta_{\rho,j}(x)$, say with use $\zeta \oplus \sigma$. At some stage $t > \mathtt{init}(\eta)[\alpha]$, x is chosen as a follower for η. Since large followers are chosen, we must have $t < s$; so η initializes weaker agents at s. η's action at s ensures that $\sigma \subset \Theta(R)$ and $\zeta \subset B$ from s onwards (this is because $s > r^{**}$). It follows that ρ will not redefine $\Delta_{\rho,j}(x)$ at any other stage after r^{**}.

Let $K'(\eta, \rho, j)$ be the set of followers $x \in K(\eta)$ such that η initializes for the sake of a $\Delta_{\rho,j}(x)$ computation after stage r^{**}. $K'(\eta, \rho, j)$ is an r.e. subset of $K(\eta)$

hence is α-finite. By Σ_2-admissibility, we know that there is a bound to the stages at which η initializes weaker agents because ρ defines a $\Delta_{\rho,j}$ computation.

Also, the set of pairs (ρ, j) for which η ever initializes is α-finite; again by Σ_2-admissibility, we know that after some stage, η stops initializing on behalf of any stronger ρ.

We have proved:

COROLLARY 3.14. *If η is any node which is treated fairly, then after some stage η doesn't initialize weaker nodes.*

Now fairness follows easily.

LEMMA 3.15. *Every node on the true path is treated fairly.*

PROOF. By induction on $\beta < \lambda$, we show that $\texttt{init}(\delta \restriction \beta)[\alpha] < \alpha$. Suppose up to β. There are two kinds of nodes which may initialize $\delta \restriction \beta$.

- Nodes $\delta \restriction \gamma$ for $\gamma < \beta$. By induction and by corollary 3.14, for each such γ, there is some stage s after which $\delta \restriction \gamma$ does not initialize weaker nodes. Let s_γ be the least such stage; $\gamma \to s_\gamma$ is $\Pi_1(J_\alpha)$ hence bounded.
- Nodes η which lie to the left of $\delta \restriction \beta$. We know that after some stage s^*, no such node is ever accessible. Thus the set of nodes which are to the left of $\delta \restriction \beta$ and are ever accessible is α-finite. After s^*, each such node may act at most once; the set of nodes which do is r.e. and so α-finite. It follows that after some stage, no node which lies to the left of $\delta \restriction \beta$ initializes other nodes. □

COROLLARY 3.16. *Every finitary-type requirement (that is, a requirement of type N, D, K) is met.*

2.3. Meeting the Requirements.

Let us first examine the positive ordering requirements.

A number y goes into H_n iff y enters $C_1^{[n]}$; so $H_n \leqslant_\alpha C_1$.

LEMMA 3.17. *For every n, $H_n \leqslant_\alpha C_0$.*

PROOF. Let ρ be the nth pregate on the true path, and let r^* be the stage which witnesses that ρ is treated fairly. Let χ be the correct guess of outcomes up to $\operatorname{dom} \rho$.

Every y which enters H_n also earlier enters $C_0^{[n]}$. On the other hand, suppose that y enters $C_0^{[n]}$ at some stage $s_0 > r^*$ (and is placed in some cage). Let s_1 be the least stage after s_0 at which ρ, χ are accessible. We claim that either y is enumerated into H_n at s_1 (or earlier), or it is never enumerated into H_n. As we can find s_0 from y using the C_0 oracle and can find s_1 from s_0 effectively, this gives us the required reduction.

Suppose then that y is not enumerated into H_n by stage s_1. If y was issued by a node to the right of ρ then it is cancelled by s_1. Otherwise, at the beginning of s_1, y is waiting in ρ's cage. If $\xi(y)$ is to the left of χ then we know that after s_1, y will never require attention and so will not end up in H_n. If $\xi(y)$ lies to the right of χ then y is cancelled by s_1. If $\xi(y)$ extends χ then at s_1, y requires attention. If it is not enumerated into H_n then it must be cancelled at s_1 (by a stronger ball, or because of failure to be promptly permitted). □

LEMMA 3.18. *For every $n > 0$, $H_n \leqslant_\alpha H_{n-1} \oplus E_{n \bmod 2}$.*

PROOF. This is classical. Whenever x is a ball on the machine targeted for H_n, x has a trace y targeted for either H_{n-1} or for $E_{n \bmod 2}$. x does not enter H_n unless y enters its target set; and there are only finitely many such traces appointed. □

Exactly as before, if $j \preccurlyeq i$, then $G_j \leqslant_\alpha G_i \oplus L$. Also, as balls entering any set need permission (either prompt or regular), all sets are recursive in U.

2.3.1. *Infima.* Let η be an agent on the true path which works for an infimum requirement $Z \in \{A_\Phi, Y_{n,\Phi}\}$. In the first case, let $X_0 = C_0, X_1 = C_1$ and $X = H_{<n(\eta)} = \oplus_{k<n(\eta)} H_k$. In the second case, let $X_0 = H_{n-1} \oplus E_{n \bmod 2}, X_1 = F_{n \bmod 2}$, and $X = H_n$.

Suppose that the hypothesis of the requirement holds: $\Phi(X_0) = \Phi(X_1)$ is total; we want to show how to calculate this function from X. Let r^* witness that η is treated fairly and let χ be the correct guess of outcomes up to $\operatorname{dom}\eta+1$. There are unboundedly many η-expansionary stages, so $\chi(\operatorname{dom}\eta) = r(\eta) = 0$ is the correct outcome.

In the analysis which follows, it is important not only at what stage various events happen but also at what phase. For notational simplicity, we refer to "moments" in time, but write down the stage only. In particular, the moment a certain computation is injured is the exact moment at which the injuring balls enter the relevant set; and the moment at which η is expansionary is the beginning of phase two of an η-expansionary stage, just when $\delta[s]$ is computed and balls are cancelled, but before balls move.

We need to better define some notions and to describe η's unique point of view, which we shall adopt in this subsection. Suppose that $Z = A_\Phi$. A ball, targeted for some H_n, residing at a gate, hole, corral or in the bin is said to be also targeted for C_0; such a ball crosses η when it is released by a gate above η and ends up residing below η. A ball, targeted for some H_n, which resides in some cage ρ, is said to be also targeted for C_1; such a ball crosses η when it is released from the gate and enumerated. In fact, η thinks of these as two different balls, the second one issued by ρ, rather than by η.

However, if $Z = Y_{n,\Phi}$, then balls lying in a cage are not considered by η to really reside in that cage; in η's eyes, they are still waiting for enumeration at the bottom of the machine, so the notion of a ball lying below η depends on Z.

Because of that, balls released from a cage above η are not considered by η to be crossing η at that stage. It follows that in either case, in η's eyes, any ball can cross η at most once.

DEFINITION. A ball z which is on the machine at stage $s > r^*$, is called *dormant* at that stage if one of the following occurs:

- z belongs to a hole which is stronger than η.
- Members of z's entourage are waiting in the bin or in some cage and have a trace for B which is not an element of B.
- Members of z's entourage are waiting at a gate ρ below η and $\xi(z)$ is to the left of $\chi \restriction \operatorname{dom}\rho$.

LEMMA 3.19. (1) *If z is dormant at s then no balls of z's entourage ever enter any sets; in fact if they ever receive attention after s it is to ask for prompt permission, at which point they are cancelled.*

(2) *If η, χ are accessible at s and z is a ball which lies below η at s and never receives attention after s, then z is dormant at s.*
(3) *The relation "z is dormant at s" is recursive in B.*
(4) *If z is dormant at s, $t > s$ and z is not already cancelled at t, then z is dormant at t.*

PROOF. (1). Let z be dormant at s. If z is issued by a hole which is stronger than η, then z cannot receive attention after s since $s > r^*$. If balls in z's entourage are waiting in the bin and have a trace for B which does not enter B then z does not receive attention after s. If balls in z's entourage are waiting in some cage and have a trace for B which does not enter B, then these balls can never receive permission; they are either cancelled, or never receive attention after s. If balls in z's entourage are waiting at some gate ρ below η, and $\xi(z)$ is to the left of $\chi \restriction \operatorname{dom} \rho$, then these balls never receive attention after s, because $\xi(x) \restriction \operatorname{dom} \rho$ is never accessible after s.

(2). Suppose that at $s > r^*$, a ball z lies below η and never receives attention after s; suppose also that η, χ are accessible at s. There are three possibilities for the status of z's entourage. z may be waiting at a hole, so belongs to a hole stronger than η. Balls of z's entourage may be waiting in the bin or in some cage on the machine; the fact that z never receives attention after s implies that the last trace for B will not enter B. Finally, balls in z's entourage may be waiting at some gate ρ, necessarily below η, and z belongs to a hole which extends η (if it belongs to a hole to the right of η then z is cancelled at s). $\xi(z)$ is at least as strong as $\chi \restriction \operatorname{dom} \rho$ since otherwise z gets cancelled at s, as its guess is discovered to be incorrect. If $\chi \restriction \operatorname{dom} \rho \subset \xi(z)$ then z requires attention at s. Thus if z never requires attention then it must be the case that $\xi(z)$ lies to the left of $\chi \restriction \operatorname{dom} \rho$. In all cases we see that z is dormant at s.

(3) and (4) are easy. □

Let $y < \alpha$. For shorthand, let, for $i < 2$, $u_i[s] = \phi(X_i, y)[s]$. Consider a moment s after r^* at which the following holds:

(1) η, χ are accessible at s.
(2) $\ell(Z)[s] > y$.
(3) For both $i < 2$, $X \restriction u_i[s]$ is correct.
(4) Every ball on the machine which lies below η at s is dormant at s.

Call such a moment *y-safe*. The set of y-safe moments is (uniformly in y) recursive in X.

LEMMA 3.20. *For every $y < \alpha$, there are unboundedly many y-safe moments.*

PROOF. Suppose that after $s_0 > r^*$, $\ell(Z) > y$ is witnessed by correct Φ computations. Let $s_1 > s_0$ be a non-deficiency stage for the construction, and suppose that s_2 is the least moment after s_1 at which η, χ is accessible. We claim that in fact no balls on the machine at s_2 ever receive attention after s_2; it will follow that all balls on the machine at s_2 which lie below η at s_2 are dormant at s_2.

Suppose that at s_2, z is a ball which lies below η. We claim that z must have been on the machine at the end of s_1, and so does not receive attention after s_1. Suppose not. As η is accessible at s_2 and z is not immediately cancelled, we know that z belongs to some node η' extending η (it cannot be appointed by some stronger node after s_1 since that would initialize η). Also, $\chi \subset \xi(z)$ since otherwise

z would be cancelled at s_2. Further, the follower x to which z belongs must have been appointed after s_1, since no followers which are on the machine at s_1 ever receive attention and so do not appoint new traces. But s_2 is the least stage after s_1 at which η, χ are accessible, and so x could not have been appointed before s_2; this is a contradiction. □

The following clears a hurdle that does not exist in classical recursion theory.

LEMMA 3.21. *For every $s > r^*$, s is η-expansionary iff $\eta \in \delta[s]$ and $r(\eta)[s] = 0$.*

PROOF. We need to show that if $s > r^*$ is a limit of η-expansionary stages then it is η-expansionary. Let s be such a stage, let $s_0 < s$ be η-expansionary ($s_0 > r^*$) and let $y < \ell(Z)[s_0]$.

As long as we can, we define a double sequence of moments $\langle t_j, s_j \rangle_{j<\omega}$: t_j is the least moment after s_j (but before s) at which one of the computations $\Phi(X_0, y)[s_j]$, $\Phi(X_1, y)[s_j]$ is injured. s_{j+1} is the least η-expansionary moment after t_j. Note that s_{j+1} is the least moment after t_j at which η is accessible and $r(\eta) = 0$. Also, $s_{j+1} < s$ because s is a limit of η-expansionary stages.

We claim that the sequence $\langle t_j, s_j \rangle$ is finite. This will imply that after some $t < s$, the computations $\Phi(X_i, y)$ are not injured, which implies that these computations still exist at s (before balls move, but when $\delta[s]$ is calculated). It follows that for every η-expansionary $s_0 < s$ we have $\ell(Z)[s_0] \leqslant \ell(Z)[s]$; as s is a limit of expansionary stages, we get that s is η-expansionary.

Let x_j be the ball which injures the relevant computation at t_j. Let x be the smallest follower such that some x_j is in x's entourage; let x_{j^*} be the last x_j which is in x's entourage. We claim that x_{j^*+1} cannot exist. Suppose that it does, and let x' be the follower to whose entourage x_{j^*+1} belongs. For some $i < 2$, $x_{j^*+1} < u_i[s_{j^*+1}]$ and so x_{j^*+1} and x' are on the machine at s_{j^*+1}.

$s_0 > r^*$ so x' is issued by a node no stronger than η. η is accessible at s_{j^*+1} so x' is really issued by a node extending η; in particular $\operatorname{dom}\eta < \operatorname{dom}\xi(x')$. Also, $r(\eta)[s_{j^*+1}] = 0$ and x' is not cancelled at s_{j^*+1} so $\xi(x')(\operatorname{dom}\eta) = 0$. This implies that x' could not have been appointed between t_{j^*} and s_{j^*+1}. x' is thus on the machine at stage t_{j^*}, and is not cancelled at t_{j^*} by x, so $x' < x$ – contradiction. □

Let s^* be a y-safe stage.

DEFINITION. A moment $t > s^*$ is *superb*, witnessed by $i < 2$ and $x < \alpha$, if the following holds at t:

(1) At t, both computations $\Phi(X_0, y)$ and $\Phi(X_1, y)$ converge.
(2) η is accessible at t and $r(\eta)[t] = 0$.
(3) The computation $\Phi(X_i, y)[t]$ is X-correct.
(4) x is a ball which crosses η at t, and will injure $\Phi(X_i, y)$.
(5) Every ball which lies below η at t is dormant.

Suppose that one of $\Phi(X_0, y)[s^*]$ or $\Phi(X_1, y)[s^*]$ is injured after s^*; say $\Phi(X_i, y)$ is injured first, by some ball x.

LEMMA 3.22. *x crosses η at some moment $t > s^*$ which is superb (witnessed by i and x).*

PROOF. x cannot belong to a node stronger than η, since $s^* > r^*$. x was on the machine at s^* since $x < u_i[s^*]$; η is accessible at s^* and so x cannot belong to a node to the right of η; thus x belongs to a node which lies above η. At s^*, x (and its entire entourage) lies above η since every ball below η at s^* is dormant.

It follows that x crosses η after s^*; let t be the moment at which x crosses η. Of course, neither computation $\Phi(X_j, y)[s^*]$ was injured before t so (1) and (3) hold. Also, χ is accessible at s^* and x is not cancelled at s^*, so $\xi(x)(\operatorname{dom}\eta) = 0$; so $r(\eta)[t] = 0$, so (2) holds. It remains to show that (5) holds.

Let z be a ball which lies below η at t. If z is stronger than x then z does not receive attention between s^* and t; it follows that z was below η at s^*, so was dormant at s^*, so it dormant at t. Of course z cannot be weaker than x. We claim that z cannot belong to x's entourage.

If x is released from a gate at t then it is clear that all balls of x's entourage which do not cross η together with x, lie above η at t. So assume that x is targeted for C_1 and is released from a cage at t. Let u be the moment at which x was enumerated into C_0. There are two cases:

- $u < s^*$. In this case, between u and t, x's entourage does not receive attention. At s^*, no balls in this entourage lie below η and so they don't lie below η at t.
- If $u > s^*$ then we know that x must have crossed η between s^* and u at some moment v. Now as η works in this case for A_Φ we know that x is the smallest ball crossing η at v (perhaps with one trace for some E_j). Every $z < x$ which is in x's entourage is left above η at v in some corral, and does not leave the corral until after t. Of course every $z > x$ which is in x's entourage at t crosses η together with x at t. □

LEMMA 3.23. *If t is superb, witnessed by i and x, and $\Phi(X_{1-i}, y)$ is injured after t, then there is another superb moment $v > t$, witnessed by $1 - i$ and some $z \leqslant x$, at which the computation $\Phi(X_{1-i}, y)$ is the same as the computation at t.*

PROOF. Let u be the moment at which x enters X_i and let z be the first ball which injures $\Phi(X_{1-i}, y)$ $[t]$.

CLAIM 3.24. z enters X_{1-i} after u, and immediately after u, z resides above η.

PROOF. We know that z is on the machine at t. If z is stronger than x then z does not receive attention between t and u; also, it cannot reside below η at t (by (5) at t). z must belong to a node above η since η is accessible at t (and $t > r^*$). Thus z resides above η at t, and so at u.

z cannot be weaker than x, since then it would be cancelled at t. Assume then that z belongs to x's entourage. We know that every ball in x's entourage either crosses η with x at t or at t lies in a corral above η. If z does not belong to x's segment at t, then it does not move until u, so the claim holds. Suppose then that z crosses η together with x.

We first claim that we cannot have $Z = Y_{n,\Phi}$. Because in that case, we know that the only way that two balls crossing η at the same time, one targeted for X_0 and the other for X_1, is if the smaller one is targeted for $X = H_n$. We know that x is not targeted for X as $\Phi(X_i, y)[t]$ is X-correct. Thus $z < x$; but $X \restriction x$ is correct at t and so z cannot enter X – impossible.

If $Z = A_\Phi$ then we know that at each time, only one ball targeted for some H_n crosses η (whether it is released by a gate or a cage). Thus $z = x$. Since x is not targeted for $X = H_{<n(\eta)}$, we know that at u, $x = z$ is retargeted for C_1 and is placed in a cage which lies above η. □

It follows that z crosses η after u; let v be the moment at which z crosses η, we show superbness. (4) holds by definition. For (2), note that z is not cancelled at t, at which η was accessible and $r(\eta) = 0$, so $\xi(z)(\operatorname{dom}\eta) = 0$. It follows that at v, η must be accessible and $r(\eta) = 0$. Lemma 3.21 implies that v is actually η-expansionary, so (1) holds at v.

For (3), we show that $X \restriction u_{1-i}[t]$ is correct at t. This is because $x < u_i[t]$ and so $X \restriction x$ is correct at t, and x eliminates larger balls at t (x being targeted for X_i implies that no immediate trace of x can be targeted for X). Of course new balls appointed are greater than $u_{1-i}[t]$.

All balls of z's entourage not crossing η together with z at v are, at v, lying in corrals above η. This is the same argument as is given at the end of the proof of lemma 3.22, with z in place of x, t in place of s^* and v in place of t. Let b be a ball which lies below η at v. It follows that b is stronger than z, and so did not receive attention between t and v. It follows that b was below η at t, and so was dormant then, and so is dormant at v. This is (5). □

It follows that the value of $\Phi(X_i, y)[s^*]$ is correct, because the sequence of superb stages given by lemmas 3.22 and 3.23 must be finite (as $z \leqslant x$, and each x can occur at most twice); at the last such stage, there is a permanent computation. η thus meets its requirement.

2.3.2. *Success of Pregates.* This is exactly as in chapter 2. Let η be the pregate on the true path working for T_n, and let r^* witness that η is treated fairly.

CLAIM 3.25. *If* $\Gamma_\eta(G_n \oplus P, y) \downarrow$ *then* $\Gamma_\eta(G_n \oplus P, y) = Q(y)$.

PROOF. Let $s > r^*$, and suppose that $\Gamma_\eta(G_n \oplus P, y) \downarrow [s]$ and that at s, agent ρ, working for $M_{\Theta,\Phi,W}$ puts y into Q. ρ cannot be stronger than η since $s > r^*$. Also, ρ cannot be to the right of η; η defined $\Gamma_\eta(y)$ at stage $t < s$ and initialized all nodes to its right at t; and new chits y are picked large. Thus $\eta \subset \rho$.

Therefore ρ puts $\gamma_n(y)[s]$ into P at s, removing the computation $\Gamma_\rho(G_n \oplus P, y)$. If the computation is defined later it must be correct. □

CLAIM 3.26. *Suppose that* $\beta < \alpha$ *and that* $\Gamma_\eta(G_n \oplus P) \restriction \beta$ *stabilizes by some* α*-finite stage. Then* $\Gamma_\eta(G_n \oplus P, \beta) \downarrow$.

PROOF. Suppose that after $t > r^*$, $\Gamma_\eta(G_n \oplus P) \restriction \beta$ is fixed. At every later stage at which η is allowed to define Γ_ρ, if $\Gamma_\rho(G_n \oplus P, \beta) \uparrow$ then it gets defined, say with use τ. At such a stage enumerate τ into a functional Ξ. Let ρ be the agent which works for K_Ξ. At some stage $t > s$, Ξ acts and so preserves $\tau \subset G_n \oplus P$. □

LEMMA 3.27. *Every* T *requirement is met.*

PROOF. For every β, $\Gamma_\eta(G_n \oplus P) \restriction \beta$ eventually stabilizes. This follows from the fact that $G_n \oplus P$ is admissible and is proved by induction. Assume up to β; the function taking $y < x$ to the stage at which the correct computation $\Gamma_\eta(G_n \oplus P, y)$ is defined is weakly recursive in $G_n \oplus P$ and so is bounded.

It follows that $\Gamma_\eta(G_n \oplus P) = Q$. As $G_n \oplus P$ is admissible, $Q \leqslant_\alpha G_n \oplus P$. □

2.3.3. *The Minimality Requirements.* Again this is exactly as in chapter 2. Let η be the agent working for $M = M_{\Theta,\Phi,W}$, and let r^* witness that η is treated fairly.

OBSERVATION 3.28. Suppose that $s > t > r^*$. Suppose that at t, (y, π) is picked as a chit for a computation $\Delta_{\eta,0}(x)$ with use $\zeta \oplus \sigma$. Suppose that at s, (y, π) is transferred as a chit for a $\Delta_{\eta,j}$ computation. Definitions of Δ computations are done only at η-expansionary stages, so $y < \ell(M)[t] < \ell(M)[s]$, so y is M-confirmed at s. Also, (y, π) is still active at s. This implies that $\phi(B \oplus W \oplus P, y)[s] = \zeta \oplus \sigma \oplus \pi$ (we use niceness of Φ here). Thus $\sigma \subset \Theta(R)[s]$. Note that once the chit is cancelled it is never active again (and will never be picked again as a new chit for $\Delta_{\eta,0}$ as we pick large chits).

LEMMA 3.29. *If η declares victory after r^* then the hypotheses of M do not hold.*

PROOF. If easy victory is declared then η preserves a discrepancy between $\Theta(R)$ and W. Suppose that at $s > r^*$, η enumerates a victorious chit y into Q. Let $\zeta \oplus \sigma \oplus \pi = \phi(B \oplus W \oplus P, y)[s]$. $\sigma \subset \Theta(R)\,[s]$ holds by observation 3.28 and is preserved by η's action at s (so if $\sigma \not\subset W$ we again get an easy win.) We claim that $\zeta \oplus \pi \subset B \oplus P$ is preserved (and so that y is not M-confirmed at α). The only obstacle can be the uses of the form $\gamma_\rho(G_n \oplus P, y)$ which η enumerates into P as it declares victory. A use of this kind is enumerated only if Γ_ρ clears y at s, thus all $\gamma_\rho(G_n \oplus P, y)$ enumerated are greater than $\operatorname{dom} \pi$. □

Assume that the hypotheses of M hold. Then $\lim_{s \to \alpha} \ell(M)\,[s] = \alpha$. It follows that there are unboundedly many M-expansionary stages.

Let $n = n(\eta)$.

CLAIM 3.30. *Let $j \leqslant n$. For all x, there is a stage after which η stops defining $\Delta_{\eta,j}(x)$.*

PROOF. At each stage s at which η defines $\Delta_{\eta,j}(x)$, say with use $\zeta \oplus \sigma$, enumerate $\zeta \oplus \theta(R; \operatorname{dom} \sigma)\,[s]$ into a functional Ξ (recall observation 3.28). The success of the agent ρ which works for K_Ξ ensures that if unboundedly many attempts at defining $\Delta_{\eta,j}(x)$ are made, then both $\zeta \subset B$ and $\theta(R; \operatorname{dom} \sigma)[s]$ for some such definition will be preserved. By observation 3.13, $\Delta_{\eta,j}(x)$ doesn't get redefined after s. □

CLAIM 3.31. $\operatorname{dom} \Delta_{\eta,0}(B \oplus W) = \alpha$.

PROOF. By induction on x, we show that $\Delta_{\eta,0}(B \oplus W, x)\downarrow$. If $x \leqslant \operatorname{dom} \Delta_{\eta,0}(B \oplus W)$, then we know that $\Delta_{\eta,0}(B \oplus W) \restriction x$ eventually stabilizes. It is enough now to show that for unboundedly many s, $\Delta_{\eta,0}(B \oplus W, x) \downarrow [s]$; by claim 3.30, after some t, η stops defining $\Delta_{\eta,0}(x)$, hence all computations $\Delta_{\eta,0}(B \oplus W, x)[s]$ for $s > t$ must be the same computation, which is permanent.

Suppose that by $s^* > r^*$, $\Delta_{\eta,0}(B \oplus W) \restriction x$ is permanent. Suppose that $t > s^*$ and $\Delta_{\eta,0}(W, x) \uparrow [t]$. Find some η-expansionary stage $s > t$ such that $\ell(M)[s]$ is large enough so that there is some $y \in \alpha^{[\eta]}$ such that $t < y < \ell(M)[s]$. Now if η didn't define $\Delta_{\eta,0}(x)$ between t and s, then y is a suitable chit (because η did not define any $\Delta_{\eta,0}$ computations between t and s), thus η would define $\Delta_{\eta,0}(x)$ at s. □

CLAIM 3.32. *Say $j \in [1, n]$. If there are unboundedly many chits which are eventually $j-1$-suitable (and never cancelled), then* $\operatorname{dom} \Delta_{\eta,j}(B \oplus W) = \alpha$.

PROOF. This is like the previous claim. Letting x, s^*, t be as above, we find an expansionary stage $s > t$ such that at s, there is a chit y which is already $j-1$-suitable and is never cancelled, such that $t < y < \ell(M)[s]$; this is possible by the assumption that there are unboundedly many such y. □

CLAIM 3.33. *Suppose that at stage s, y is an active chit for a failed $\Delta_{\eta,j}$ computation. Suppose that $\rho \subset \eta$ works for T_j. Then Γ_ρ clears y at s.*

PROOF. We show that if $\Gamma_\rho(G_j \oplus P, y) \downarrow [s]$ then this computation must have been defined after the stage $t < s$ at which the chit was originally picked as a chit for a computation $\Delta_{\eta,j}(x)$. Obviously (by the size requirement for y) we have $x \leqslant y$. As the computation is failed at s, x enters G_j at a stage $u \in (t, s)$; now $x \leqslant y \leqslant \gamma_\rho(G_j \oplus P, y)\,[u]$ (if $\Gamma_\rho(G_j \oplus P, y) \downarrow [u]$), destroying the latter computation. □

LEMMA 3.34. *There cannot be unboundedly many chits which are eventually n-suitable and which are never cancelled.*

PROOF. We show that if there are unboundedly many such chits then η declares victory after r^*.

By assumption we can find unboundedly many $y > \mathtt{Rest}(\eta)$ and a stage s at which η is accessible, $r(\eta)[s] = 0$ and $\mathtt{Rest}(\eta)\,[s] = \mathtt{Rest}(\eta)$, such that at s, y is an active chit for a failed $\Delta_{\eta,n}$ computation and such that $y < \ell(M)[s]$. Then y is victorious at s, and at s, η may win by enumerating y into Q, if y is permitted promptly by U. The set of such ys is r.e. and unbounded and so one will be permitted. □

As in chapter 2, the following concludes the verifications.

LEMMA 3.35. *Suppose that $j \leqslant n$ and $\operatorname{dom} \Delta_{\eta,j}(B \oplus W) = \alpha$ but $\Delta_{\eta,j}(B \oplus W) \neq^* G_j$. Then there are unboundedly many chits which are eventually j-suitable and are never cancelled.*

PROOF. Much of the proof goes along classical lines. Fix $\beta < \alpha$. Let the functional Ξ converge at $t > r^*$ with use $\zeta \oplus \rho \oplus \pi \subset B \oplus R \oplus P\,[t]$ if there is some $\sigma \subset W[t]$ such that $\sigma \subset \Theta(\rho)[t]$ and there is some $y > \beta$ such that at t, (y, π) is an active chit for a failed $\Delta_{\eta,j}$ computation with use $\zeta \oplus \sigma$. If Ξ converges unboundedly often, the success of K_Ξ would show that there is some j-suitable chit which is never cancelled and which is greater than β (protection of $\zeta \oplus \rho \oplus \pi \subset B \oplus R \oplus P$ ensures that $\sigma \subset W$).

Suppose that $t^* > r^*$ is any stage. Take some $x > \beta, t^*$ such that $\Delta_{\eta,j}(W, x) \neq Q(x)$. Suppose that the correct $\Delta_{\eta,j}(x)$ computation is defined at stage $s > t^*$ with associated chit (y, π). Let $t > s$ be the stage at which this computation fails; some agent ρ enumerates x into G_j at t. The standard argument shows that $\eta \subset \rho$; ρ cannot be stronger than η as $t > r^*$, and η cannot lie to the right of ρ since η initializes nodes to its right at s and $s > x$. This argument shows that x is a follower for ρ at s and so that ρ initializes weaker requirements on η's behalf at s.

Let u be the least stage greater than t at which $r(\eta) = 0$. The key is noticing that u is M-expansionary, as it cannot be a limit of M-expansionary stages. Thus $y < \ell(M)[u]$.

We verify that $\pi \subset P$ is preserved until u. As mentioned, ρ initializes weaker agents at s. ρ is not initialized until at least after t, as x is still a follower at t. Thus $\pi \subset P$ is preserved between s and t by ρ's action. Between t and u, restraint which is imposed by η protects $\pi \subset P$; recall that s is an M-expansionary stage.

Now at u, y is M-confirmed so $\sigma \subset \Theta(R)[u]$ (note that $\sigma \subset W$ is always true since we picked the correct $\Delta_{\eta,j}(B \oplus W)$ computation). Thus $\Xi(B \oplus R \oplus P) \downarrow [u]$. $\square$

CHAPTER 4

A Negative Result Concerning Effective Successor Models

We prove quite the opposite of the result of chapter 3 for a wide class of admissible ordinals α.

1. Preparation: Some Complexity Calculations

Let U be an amenable set. We calculate the complexities of some relations; let $\langle \Phi_e \rangle$ be an effective list of all nice functionals.

The relation

$$\sigma \subset U$$

for a string σ is $\Delta_0(J_\alpha, U)$. The relation

$$\sigma \subset \Phi_e(U)$$

is $\Sigma_1(J_\alpha, U)$. The relation

$$\Phi_e(U) \text{ is total}$$

is $\Pi_2(J_\alpha, U)$; we call this relation (set) $\mathtt{Tot}(U)$. Note that

$$\{\Phi_e(U) \,:\, e \in \mathtt{Tot}(U)\}$$

only varies over all *amenable* sets recursive in U; we don't mind as we only need r.e. degrees.

A *U-index* is an element of $\mathtt{Tot}(U)$. For shorthand, we write X_e instead of $\Phi_e(U)$ for all $e \in \mathtt{Tot}(U)$. We consider the following relations among U-indices:

The relation

$$\Phi_e(X_a) = X_b$$

is $\Pi_2(J_\alpha, U)$ because $\sigma \subset \Phi_e(\Phi_i(U))$ is $\Sigma_1(J_\alpha, U)$. Thus $X_a \leqslant_\alpha X_b$ and $X_a \equiv_\alpha X_b$ are $\Sigma_3(J_\alpha, U)$.

The relation

$$X_a = W_e$$

is $\Pi_2(J_\alpha, U)$. Thus $X_a \equiv_\alpha W_e$ and "X_a has r.e. degree" are $\Sigma_3(J_\alpha, U)$.

There is a recursive function $\mathtt{join}$ such that for all $a, b \in \mathtt{Tot}(U)$, $X_{\mathtt{join}(a,b)} = X_a \oplus X_b$.

LEMMA 4.1. *Suppose that $A(x, y)$ is $\Sigma_n(J_\alpha, U)$ $(n > 0)$. Then the formula*

$$\text{“}t \textit{ is finite } \ \& \ \forall x \in t \; A(x, y)\text{”}$$

is also $\Sigma_n(J_\alpha, U)$.

PROOF. We first prove this for $n = 1$. This is fairly easy. Suppose that C is a bounded formula such that $A(x, y) \Leftrightarrow \exists z\ C(x, y, z)$. Then the required formula is equivalent to

$$\text{"}t \text{ is finite } \ \& \ \exists u\, \forall x \in t\, \exists z \in u\ C(x, y, z)\text{"};$$

this is because J_α is closed under taking finite subsets.

Suppose by induction that the statement holds for n. Let $A(x, y)$ be a formula which is $\Sigma_{n+1}(J_\alpha, U)$ and let C be the $\Pi_n(J_\alpha, U)$ formula such that for all x and y, $A(x, y) \iff \exists z C(x, y, z)$. Let $\psi(x, y, u)$ be the formula

$$\text{"if } u \text{ is finite then } \exists z \in u\ C(x, y, z)\text{"}.$$

By the induction hypothesis, ψ is $\Pi_n(J_\alpha, U)$. By quantifier contraction,

$$\forall x \in t\, \psi(x, y, u)$$

is also $\Pi_n(J_\alpha, U)$. Again because J_α is closed under taking finite subsets, the required formula is equivalent to

$$\text{"}t \text{ is finite } \ \& \ \exists u\ (u \text{ is finite } \ \& \ \forall x \in t\ \psi(x, y, u))\text{"}$$

which is easily seen to be $\Sigma_{n+1}(J_\alpha, U)$. □

2. More on Effective Models

Recall that the correctness condition $\chi_{\mathtt{effective}}(\bar{\mathbf{p}}, \bar{\mathbf{e}})$, defined in the beginning of chapter 3 states that $\bar{\mathbf{p}}$ codes an effective successor model of arithmetic, witnessed by $\bar{\mathbf{e}}$. We note that if $\chi_{\mathtt{effective}}(\bar{\mathbf{p}}, \bar{\mathbf{e}})$ holds, then the function taking $\mathbf{x} \in M_{\bar{\mathbf{p}}}$ to $\sum_{\bar{\mathbf{p}}} \mathbf{x}$ is definable from $\bar{\mathbf{p}}$ (uniformly in such $\bar{\mathbf{p}}$).

Fix $\bar{\mathbf{p}}, \bar{\mathbf{e}}$ satisfying $\chi_{\mathtt{effective}}$. Let U be an amenable r.e. set which computes all the parameters $\bar{\mathbf{p}}, \bar{\mathbf{e}}$. Using the notation of the first subsection, consider the following relations.

- "$X_a \in 0^{M_{\bar{\mathbf{p}}}}$" is $\Sigma_3(J_\alpha, U)$ (as we can use a particular U-index for an element of $0^{M_{\bar{\mathbf{p}}}}$).
- There is a $\Sigma_3(J_\alpha, U)$ relation S such that for all a, if $\deg(X_a) \in M_{\bar{\mathbf{p}}}$ then for all $b \in \mathtt{Tot}(U)$, $S(a, b)$ holds iff $\deg(X_b) = (\deg(X_a) + 1)^{M_{\bar{\mathbf{p}}}}$. S is the following relation:

$$X_b \geqslant_\alpha \mathbf{b} \ \& \ X_b \leqslant_\alpha \mathbf{r} \ \& \ \mathbf{q} \leqslant_\alpha X_b \vee \mathbf{p} \ \& \ \bigvee_{i<2} (X_b \leqslant_\alpha X_a \vee \mathbf{e}_i \ \& \ X_b \leqslant_\alpha \mathbf{f}_i)\,.$$

- "$X_a \in n^{M_{\bar{\mathbf{p}}}}$" (for $n < \omega$) is $\Sigma_3(J_\alpha, U)$. It holds iff there is a string σ of length $n + 1$ such that $X_{\sigma(0)} \in 0^{M_{\bar{\mathbf{p}}}}$ and for all $i < n$, $S(\sigma(i), \sigma(i + 1))$. This uses lemma 4.1.

We now add our assumption on U and α which is: every $\Sigma_3(J_\alpha, U)$ relation with domain ω is uniformized by some α-finite function.

Let $\psi(n, (a, e))$ be the relation

$$\text{"}a \in \mathtt{Tot}(U) \ \& \ X_a \in n^{M_{\bar{\mathbf{p}}}} \ \& \ X_a = W_e\text{"}.$$

ψ is $\Sigma_3(J_\alpha, U)$. By assumption, there is an α-finite function $f\colon \omega \to \alpha^2$ which uniformizes ψ. Let

$$C = \bigoplus_{n<\omega} W_{(f(n))_1} = \bigoplus_{n<\omega} X_{(f(n))_0}.$$

C is r.e. and recursive in U.

LEMMA 4.2. *In the r.e. degrees below U, C is the least upper bound for the standard part of $M_{\bar{\mathbf{p}}}$ ($\{n^{M_{\bar{\mathbf{p}}}} : n < \omega\}$).*

PROOF. Suppose that $Y \leqslant_\alpha U$ is amenable and that for all n, $n^{M_{\bar{\mathbf{p}}}} \leqslant_\alpha Y$. Let $R(n, e)$ be the relation $\Phi_e(Y) = X_{(f(n)_1)}$. This is $\Sigma_3(J_\alpha, U)$; again by assumption we get an α-finite function $g \colon \omega \to \alpha$ uniformizing R; g gives us a way to uniformly compute C from Y. □

LEMMA 4.3. *If $\mathbf{x}$ belongs to the nonstandard part of $M_{\bar{\mathbf{p}}}$ then $\sum_{\bar{\mathbf{p}}} \mathbf{x} \not\leqslant_\alpha C$.*

PROOF. For every nonstandard $\mathbf{y} \in M_{\bar{\mathbf{p}}}$ and every $n < \omega$, $n^{M_{\bar{\mathbf{p}}}} \leqslant \sum_{\bar{\mathbf{p}}} \mathbf{y}$ and so $C \leqslant_\alpha \sum_{\bar{\mathbf{p}}} \mathbf{y}$. Suppose that $\mathbf{x} \in M_{\bar{\mathbf{p}}}$ is nonstandard and assume for contradiction that $\sum_{\bar{\mathbf{p}}} \mathbf{x} \leqslant_\alpha C$, so we have $\sum_{\bar{\mathbf{p}}} \mathbf{x} \equiv_\alpha C$. $\mathbf{x}$ is nonstandard so there is some $\mathbf{y} <^{M_{\bar{\mathbf{p}}}} \mathbf{x}$ which is also nonstandard; $\sum_{\bar{\mathbf{p}}} \mathbf{y} \leqslant \sum_{\bar{\mathbf{p}}} \mathbf{x}$ so we have $\sum_{\bar{\mathbf{p}}} \mathbf{y} = \sum_{\bar{\mathbf{p}}} \mathbf{x}$. This is impossible because $\langle \sum_{\bar{\mathbf{p}}} \mathbf{y} : \mathbf{y} \in M_{\bar{\mathbf{p}}} \rangle$ is strictly increasing with $<^{M_{\bar{\mathbf{p}}}}$ (for example, $\mathbf{y} \leqslant \sum_{\bar{\mathbf{p}}} \mathbf{x}$ and $\mathbf{y} \not\leqslant \sum_{\bar{\mathbf{p}}} \mathbf{y}$.) This is a contradiction. □

Let $\theta(x, \mathbf{c}, \bar{\mathbf{p}})$ state that $x \in M_{\bar{\mathbf{p}}}$ and $\sum_{\bar{\mathbf{p}}} x \leqslant \mathbf{c}$. We thus see that $\theta(\mathcal{R}_\alpha, \mathbf{c}, \bar{\mathbf{p}})$ is the standard part of $M_{\bar{\mathbf{p}}}$. We have shown:

THEOREM 4.4. *Suppose that α is admissible U is an amenable r.e. set and that every $\Sigma_3(J_\alpha, U)$ relation with domain ω is uniformized by an α-finite function. Then for all $\bar{\mathbf{p}}, \bar{\mathbf{e}}$ below U satisfying $\chi_{\texttt{effective}}$, there is some $\mathbf{c} \leqslant_\alpha U$ which is the least upper bound in the degrees below U of the standard part of $M_{\bar{\mathbf{p}}}$ and moreover that standard part is definable as $\theta(\mathcal{R}_\alpha, \mathbf{c}, \bar{\mathbf{p}})$.*

We remark again that for the αs discussed in chapter 3, the models constructed show that the above statement does not hold for any promptly simple U.

3. Examples of α and U

We give four examples for pairs α and U satisfying the above assumption. In all cases we of course must have $\mathrm{cf}_{\Sigma_3(J_\alpha)}(\alpha) > \omega$.

3.1. $\varrho^2_\alpha = \alpha$ and U is low. In this case, $\Sigma_3(J_\alpha, U) = \Sigma_3(J_\alpha)$, and every $\Sigma_3(J_\alpha)$ function from ω to α is α-finite.

Let C be a $\Sigma_3(J_\alpha, U)$-relation with domain ω. U is low so U is admissible and so $\Sigma_2(J_\alpha, U)$ is the same as r.e. in U' which is the same as r.e. in $0'$ which is the same as Σ_2. By Jensen's uniformization theorem, C can be uniformized by some Σ_3 function $f \colon \omega \to \alpha$.

Let A be a $\Pi_2(J_\alpha)$ relation such that $f(n) = \gamma$ iff $\exists z\, A(n, \gamma, z)$. Consider A as a relation between n and (γ, z); by Jensen's uniformization theorem there is a function g which is $\Sigma_3(J_\alpha)$ and uniformizes A. As by assumption $\mathrm{cf}_{\Sigma_3(J_\alpha)}(\alpha) > \omega$, it follows that there is some α-finite $\beta > \sup \operatorname{range} f$ such that for all $n < \omega$ there is some $z < \beta$ such that $A(n, f(n), z)$ holds. $A \cap (\omega \times \beta^2)$ is α-finite by assumption, and f can be effectively defined by that set.

3.2. α is Σ_2-admissible and U is low$_2$. In this case too we have $\Sigma_3(J_\alpha, U) = \Sigma_3(J_\alpha)$ and every $\Sigma_3(J_\alpha)$ function from ω to α is α-finite.

Let $V = U'$. We work in the admissible structure $M = (J_\alpha, 0')$. V is low in the M-degrees; again we get that $\Sigma_2(M, V) = \Sigma_2(M)$. But $\Sigma_2(M) = \Sigma_3(J_\alpha)$ and $\Sigma_2(M, V) = \Sigma_3(J_\alpha, U)$ because $\Sigma_2(J_\alpha, U) = \Sigma_1(J_\alpha, V)$ (by U's admissibility) and the latter is the same as $\Sigma_1(M, V)$ as $0' \leqslant_\alpha V$.

Let C be a $\Sigma_3(J_\alpha, U)$ relation with domain ω. So C is $\Sigma_3(J_\alpha)$ and so can be uniformized by some $\Sigma_3(J_\alpha)$ function $f\colon \omega \to \alpha$.

Again work in M. f is $\Sigma_2(M)$. Given any $f\colon \omega \to \alpha$ which is $\Sigma_2(M)$ (and so actually $\Delta_2(M)$, we get an M-recursive approximation $f[s]$ to f; by the fact that $\mathrm{cf}_{\Sigma_2(M)})(\alpha) > \omega$ we know that the approximation has to stabilize by some α-finite stage, so $f[s] = f$ for some $s < \alpha$; by admissibility of M, $f[s] \in J_\alpha$.

3.3. α is Σ_3-admissible, $\mathrm{cf}_{\Sigma_4(J_\alpha)}(\alpha) > \omega$ and U is in any r.e. degree. In this case every $\Sigma_3(J_\alpha, U)$ relation is $\Sigma_4(J_\alpha)$ relation and so can be uniformized by a $\Sigma_4(J_\alpha)$ function. Now if $f\colon \omega \to \alpha$ is $\Sigma_4(J_\alpha)$ then it is $\Sigma_2(M)$ where $M = (J_\alpha, 0'')$ is admissible and as in the previous subsection we get that f is α-finite.

3.4. $\varrho^3_\alpha = \alpha$, $\mathrm{cf}_{\Sigma_4(J_\alpha)}(\alpha) > \omega$ and U is in any r.e. degree. Again, $\Sigma_3(J_\alpha, U)$ is Σ_4 and every Σ_4 function $f\colon \omega \to \alpha$ is α-finite; this is as in subsection 3.1.

CHAPTER 5

A Nonembedding Result

We show that lattices with critical triples (for example, the 1-3-1) cannot be embedded below a degree which cannot compute a cofinal ω-sequence in α.

DEFINITION. Let $A \subset \alpha$ be amenable and let Ψ be a nice functional. $\gamma < \operatorname{dom}\Psi(A)$ is a Ψ, A*-closure point* if for all $\beta < \gamma$, $\psi(A;\beta) < \gamma$.

PROPOSITION 5.1. *Let A be amenable, Ψ a nice functional, and suppose that $\operatorname{dom}\Psi(A) = \alpha$. Also assume that $\operatorname{rcf}(A) > \omega$. Then the collection of Ψ, A-closure points is a club of α which is recursive in A.*

PROOF. It is immediate from the definition that the collection of closure points is closed. The function $\gamma \to \psi(A;\gamma)$ is weakly recursive in A. For all $\gamma < \alpha$, the sequence $\langle \gamma_n \rangle_{n<\omega}$ defined inductively by $\gamma_0 = \gamma$, $\gamma_{n+1} = \psi(A;\gamma_n + 1)$ is strictly increasing and weakly recursive in A; hence, bounded below α. Its supremum is a Ψ, A-closure point. □

Since we'd like to get points which are both Ψ_0, A- and Ψ_1, A-closure points, we mention

PROPOSITION 5.2. *Suppose that C, D are clubs of α and that $\operatorname{rcf}(C \oplus D) > \omega$. then $C \cap D$ is a club.* □

If A is r.e., then we can guess, at some stage $s < \alpha$, whether some γ is a closure point: we say that $\gamma < \operatorname{dom}\Psi(A)\,[s]$ is a Ψ, A-closure point *at s* if for all $\beta < \gamma$, $\psi(A;\beta)[s] < \gamma$.

OBSERVATION 5.3. Suppose A is r.e. and amenable, and that γ is a Ψ, A-closure point. There is some $s < \alpha$ such that $A \restriction \psi(A;\gamma) = A[s] \restriction \psi(A;\gamma)$. Then for all $t > s$, γ is a Ψ, A-closure point at t, detected by the correct computations.

DEFINITION (Weinstein [**Wei88**]). Suppose L is an upper semi-lattice. A *weak critical triple* in L is a triple a_0, a_1, b such that $a_0 \vee b = a_1 \vee b$, $a_0 \not\leqslant b$ and there is no $e \leqslant a_0, a_1$ such that $a_0 \leqslant e \vee b$.

Suppose that A_0, A_1 and B are amenable and r.e., and that $A_0 \oplus B \equiv_\alpha A_1 \oplus B$. Then there are nice functionals Ψ_0 and Ψ_1 such that $\Psi_0(B \oplus A_1) = A_0$ and $\Psi_1(B \oplus A_0) = A_1$. Let $A = B \oplus A_0 \oplus A_1$. Assume that $\operatorname{rcf}(A) > \omega$.

We construct an r.e. set E. For all $\gamma < \alpha$ and $s < \alpha$ we'll have $E^{[\gamma]}[s] \in \alpha$. The instructions are: Suppose that at stage s, γ is both a $\Psi_0, B \oplus A_0$ and a $\Psi_1, B \oplus A_1$-closure point, and that a change in A_0 or A_1 is detected below γ (that is, $A_i \restriction \gamma[s] \neq A_i \restriction \gamma[t]$ for all $t < s$ for i either 0 or 1.) It follows that either

$B \upharpoonright \gamma[s] \neq B \upharpoonright \gamma$ or $A_{1-i} \upharpoonright \gamma[s] \neq A_{1-i} \upharpoonright \gamma$. We can effectively find some stage $t > s$ at which some $x < \gamma$ enters either B or A_{1-i}.

Say that we found a change at A_{1-i} below γ. We then enumerate $E^{[\gamma]}[s]$ into $E^{[\gamma]}$ at s.

By permitting (and the fact that $\gamma \leqslant \langle \gamma, i \rangle$ for any i), $E \leqslant_\alpha A_0, A_1$.

We give an algorithm to compute both A_i from $E \oplus B$. At stage s, suppose that γ is both a $\Psi_0, B \oplus A_0$ and a $\Psi_1, B \oplus A_1$-closure point and that $B \upharpoonright \gamma[s]$ is correct. Then $A_i \upharpoonright \gamma[s]$ are correct iff $E^{[\gamma]}[s] \notin E^{[\gamma]}$. All of A_i can be thus computed, because there are unboundedly many γ which are both $\Psi_0, B \oplus A_0$ and $\Psi_1, B \oplus A_1$-closure points (and that fact is eventually approximated, witnessed by the correct computations).

Thus we can conclude:

THEOREM 5.4. *If* $\operatorname{rcf}(\mathbf{a}) > \omega$ *then in the* α*-r.e. degrees, there is no weak critical triple below* $\mathbf{a}$.

CHAPTER 6

Embedding the 1-3-1 Lattice

This chapter complements chapter 5 by proving:

THEOREM 6.1. *If D is a collapsible r.e. set, then there is an embedding of the 1-3-1 lattice in the α-r.e. degrees below D.*

For the notion of a collapsible set (or degree) see appendix D.

For the construction, we use ideas for working below nonlow$_2$ degrees from [**DS96**] and Fejer's construction of a branching degree ([**Fej82**]).

1. Preparation

Let D be an amenable r.e. set such that there is some bijection $p\colon \omega \Longleftrightarrow \alpha$, weakly recursive in D.

LEMMA 6.2. *There is a recursive approximation $p[s]$ for p with the following properties:*

- *For every s there is some $n < \omega$ such that $\neg\,(p(n)[s] = \lim_{t\to s} p(n)[t])$.*
- *There is an increasing sequence $\langle \beta_n \rangle$, weakly recursive in D, such that after β_n, $p \restriction n[s]$ is permanent.*

PROOF. Let $p_0[s]$ be a recursive approximation for p which we get by approximating p's computation from D. Define $p[s]$ by induction on s. If $\lim_{t\to s} p(n)[t]$ does not exist for some $n < \omega$ then let $p[s] = p_0[s]$. Otherwise, let $p[s]$ be $p_0[u]$ for the least $u > s$ such that for some n, $p_0(n)[u] \neq \lim_{t\to s} p(n)[t]$. If $p_0 \restriction n[s]$ is correct then so is $p \restriction n$. $\square$

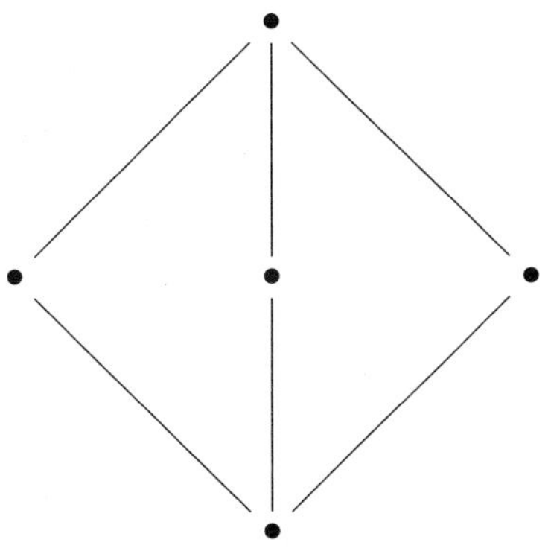

FIGURE 6.1. The 1-3-1 lattice

$\langle \beta_n \rangle$ is necessarily cofinal in α. For $s < \omega$ we let $n(s)$ be the least n such that either $\lim_{t\to s} p(n)[t]$ does not exist, or exists and is not equal to $p(n)[s]$. For all $s \geqslant \beta_k$ we have $n(s) \geqslant k$.

PROPOSITION 6.3. *For every $f\colon \omega \to \alpha$ which is weakly recursive in $0'$, there is some $g\colon \omega \to \alpha$, weakly recursive in D, which is not dominated by f, i.e.*

$$\{n < \omega \,:\, g(n) \geqslant f(n)\}$$

is infinite.

PROOF. By Shore ([**Sho76b**]), $D' \equiv_\alpha 0''$.

Consider the admissible collapse above D (see appendix D). Let $\mathcal{D}$ be D's collapse and let $\mathcal{K} \geqslant_T \mathcal{D}$ be $0'$'s collapse. Theorem D.11 implies that $\mathcal{K}'' \equiv_T \mathcal{D}''$. In other words, $\mathcal{K} \in \mathbf{L}_2(\mathcal{D})$. In particular, $\mathcal{K}$ is not high over $\mathcal{D}$. Thus, for every $f \leqslant_T \mathcal{K}$ there is some $g \leqslant_T \mathcal{D}$ which is not dominated by f.

By the definition of the admissible collapse, we have, for every $f\colon \omega \to \omega$ which is weakly recursive in $0'$, some $g\colon \omega \to \omega$, not dominated by f, which is weakly recursive in D.

Let $f\colon \omega \to \alpha$ be weakly recursive in $0'$. For $k < \omega$ let $\tilde{f}(k)$ be the least n such that $f(k) < \beta_n$. $\tilde{f}$ is weakly recursive in $0'$, so there is some $\tilde{g}\colon \omega \to \omega$, weakly recursive in D, which is not dominated by $\tilde{f}$. Let $g(k) = \beta_{\tilde{g}(k)}$; g is as required. □

PROPOSITION 6.4. *There is a function $f\colon \omega \to \alpha$, weakly recursive in $0'$, such that for every r.e. set W, if for all $n < \omega$, $W^{[n]} \setminus \beta_n$ is finite, then for all but finitely many $n < \omega$, $f(n) > \max W^{[n]}$.*

PROOF. The set $\{e \,:\, W_e \text{ is infinite}\}$ is r.e. This is because every infinite r.e. set W contains an infinite, α-finite r.e. set, which is enumerated into W by an α-finite stage. Thus this set is recursive in $0'$. Given e such that W_e is finite, $0'$ can compute $\max W_e$ (that is to say, the function $e \to \max W_e$, defined on all e such that W_e is finite, is weakly recursive in $0'$.)

Thus the function $m(e,n)$, which returns $\max W_e^{[n]} \setminus \beta_n$ if $W_e^{[n]} \setminus \beta_n$ is finite and nonempty, and β_n otherwise, is weakly recursive in $0'$. So we can let

$$f(n) \;=\; \max\{m(p(k),l) \,:\, k,l \leqslant n\}$$ □

Let $g\colon \omega \to \alpha$ be weakly recursive in D, which is not dominated by f (of proposition 6.4). Let $g[s]$ be a recursive approximation for g which we get from approximating g's computation from D. For every n, D can effectively find a stage after which $g \restriction n[s]$ is permanent. We may assume that $\sup \operatorname{range} g[s] \leqslant s$.

2. The Embedding

We construct A_0, A_1, A_2 and B fulfilling the requirements:

P: $\operatorname{otp} \neg B = \omega$.
$H_{\Psi,i}$: $\Psi(B) \neq A_i$.
$G_{\Phi,i,j}$: If $i \neq j$ and $\Phi(B \oplus A_i) = \Phi(B \oplus A_j) = C$ then $C \leqslant_\alpha B$.

and such that $A_i \leqslant_\alpha B \oplus A_j \oplus A_k$, and $B, A_i \leqslant_\alpha D$.

2.1. Elements of the Construction. All functionals given are nice. At stage $s < \alpha$, we enumerate $\neg B$ as $\langle b_i[s]\rangle_{i<\alpha}$ (At the end of the construction, enumerate $\neg B$ as $\langle b_n\rangle_{n<\omega}$.) As we expect to have $\operatorname{otp}\neg B = \omega$, we only consider computations which mention finitely much negative information about B. That is, we only regard computation with use bounded below $b_\omega[s]$.

Whenever a number $b_n[s]$ enters B, we also enumerate all of $\{b_m[s] : m > n\}$ into B; we may do this because we just realized that our guess about the value for each such b_m was incorrect.

We order H and G requirements effectively in order-type α. The tree of strategies will be $2^{<\omega}$. At stage s, we declare that agents of level n work for requirement $p(n)[s]$.

NOTATION. Throughout, we denote a node working for an H requirement by the letter η (and sometimes call it a "hole"). A node working for a G requirement is called a "gate" and is denoted by ρ.

CONVENTION. At the beginning of stage s, all numbers involved in the construction are below ωs. At stage s we add finitely many elements to the construction (balls, markers) which are "large", these will be chosen from $(\omega s, \omega(s+1))$; this will ensure this convention holds at $s+1$. Also note that at stage s we do *not* choose ωs to be any new ball or marker; this is not very significant but will help us show that $\operatorname{otp}\neg B \geqslant \omega$.

2.1.1. *The basic strategy for P.* P is a global "dumping" requirement. At every stage s, P enumerates all but finitely many numbers $< \omega s$ into B.

2.1.2. *The basic strategy for H.* Holes and Gates follow Lachlan's ideas for embedding the 1-3-1 in the ω-r.e. degrees, modified to use Fejer's method of constructing branching degrees. The construction is a pinball machine on the tree of agents; all followers and traces are balls on the machine, which can reside at various nodes.

A hole η working for $H_{\Psi,i}$ appoints *a follower* x and waits for it to be realized, i.e. for $\Psi(B,x)\downarrow= 0$. Positive order requirements are met by a tracing procedure. Any ball y_0 targeted for A_i is appointed a trace y_1 which is targeted for either A_j or A_k, where $\{i,j,k\} = 3$. The trace y_1 receives a trace of its own, a stage after being appointed. As in chapter 2, the collection of balls consisting of a follower and closed under taking traces, is called an *entourage*.

Once a follower x for η is released from its hole, its entire entourage moves down the tree, to the next *η-gate*. An η-gate is a gate ρ such that $\rho^\frown 0 \subset \eta$ (here 0 denotes the infinitary outcome). At the next stage at which η is accessible, the *ρ-permissible tail* of x's entourage is released from ρ and descends to the next η-gate. The ρ-permissible tail is the longest final segment of the entourage which does not contain both a ball targeted for A_i and a ball targeted for A_j, where ρ works for $G_{\Phi,i,j}$. This procedure repeats itself; each time η is accessible, there is a final segment of x's entourage, waiting at some η-gate ρ (the rest of the entourage lies at higher gates), and the ρ-permissible tail descends further to the next η-gate.

An entourage tail which is released from the lowest η-gate is placed in the *permitting bin*. After balls in the tail are permitted to enter the sets for which they are targeted, the new tail of the entourage is residing at some η-gate ρ, and the process repeats itself.

Once a follower x is enumerated into its target set A_i, η declares victory and ceases all action (until the next time it is initialized).

This orderly picture relies upon the fact that there is always a nonempty ρ-permissible tail; this is ensured by the entourage being *finite*. Thus at the beginning of every stage we winnow out entourages and make them all finite. This winnowing is coded into B, and so does not interfere with the tracing procedure.

When appointing new traces to the end of an entourage residing at a gate ρ, we target the new traces such that the ρ-permissible tail of the entourage does not get shorter. This implies that whenever some tail of the entourage leaves ρ, at least one ball in that tail originated above ρ and arrived at ρ at some point, and was not appointed as a trace at ρ. We thus have

LEMMA 6.5. *For any follower x there are only finitely many stages at which balls from x's entourage move.*

2.1.3. *The basic strategy for G.* A gate ρ builds a functional Γ_ρ with intended oracle B. The intention is that $\Gamma_\rho(B,n) = \Phi(B \oplus A_i; b_n)$, so if there is agreement between $\Phi(B \oplus A_i; b_n)$ and $\Phi(B \oplus A_j; b_n)$ at a certain stage, ρ would like to set this common value as $\Gamma_\rho(B,n)$. When ρ defines $\Gamma_\rho(B,n)$, the use will always be $(B \restriction b_m + 1)[s]$ for some $m > n$. That number $b_m[s]$ is called the *marker* for this computation and is (sloppily) denoted by $\gamma_\rho(n)$. When instructed to *remove* the computation $\Gamma_\rho(B,n)[s]$, ρ enumerates $\gamma_\rho(n)[s]$ into B (in this way ρ will be able to remove computations that are seen to be incorrect). Of course we also require that $\gamma_\rho(n+1) > \gamma_\rho(n)$ so $\operatorname{dom}\Gamma_\rho(B)[s]$ is closed downwards; at s we only define $\Gamma_\rho(B, \operatorname{dom}\Gamma_\rho(B)[s])$.

DEFINITION. If ρ works for $G_{\Phi,i,j}$, then a computation $\Gamma_\rho(B,n) = \sigma$ is *semi-believable* (at stage s) if $\operatorname{dom}\sigma = b_n\,[s]$ [if a computation is not even semi-believable, then it has no hope.] It is *l-believable* (for $l \in \{i,j\}$) if it is semi-believable and $\Phi(B \oplus A_l; b_n) = \sigma\,[s]$. It is *believable* if it is either i-believable or j-believable; it is *doubly believable* if it is both i- and j-believable.

At every stage, after every time numbers enter sets, all gates ρ on the tree examine their $\Gamma_\rho(B)$ and search for computations which are not believable ("unbelievable" computations). If one is found, it is to be removed.

While removing computations, numbers go into B, rendering even more computations unbelievable. These too will have to be removed immediately. And so on, until only believable computations are left. This process is called *cascading*.

To maintain order, when a gate ρ acts and removes computations, it initializes all nodes to the right of $\rho^\frown 0$. ρ does not initialize nodes which extend $\rho^\frown 0$, Since if the correct outcome is 0 (agreement), ρ will remove computations unboundedly often.

We remark that the dumping requirement ensures that after acting, for every gate ρ, $\operatorname{dom}\Gamma_\rho(B)[s]$ is finite.

2.1.4. *Permission.* It may be that a tail of an entourage of a follower for hole η will wait in the permitting bin eternally. To counter this, as usual, η will appoint more followers. When all followers have entourage tails in the permitting bin, η appoints a new follower. The dumping requirement P will ensure that at every stage, a hole has only finitely many followers on the machine.

The priority ordering between followers is determined by date of birth (which is the same as by size, as we appoint large followers).

We say that a number m is *permitted at* s if for some $k \leqslant m$,

$$\neg \left(g(k)[s] = \lim_{t \to s} g(k)[t] \right)$$

(in particular if the limit does not exist.) A tail of the entourage for the mth follower (in strength) of η is permitted at s exactly when m is permitted. Note that when permission is obtained, the balls must be enumerated immediately, even if η is not accessible.

2.1.5. *Initialization.* We explained when gates initialize other nodes. Holes also initialize other nodes; whenever a hole, or one of its followers, receives attention (which includes appointing a follower, some part of an entourage of a follower moves or gets enumerated, or some other special delicacy I'm reserving for later), the hole initializes all weaker nodes: nodes above and to the right (this kind of action is finitary).

Also as usual, the accessible nodes initialize nodes to the right.

Now what happens when a node is initialized? If it is a hole, then all of its followers are removed from the tree, and if it declared victory before, it is restored to the hungry state of dissatisfaction. If it is a gate, then all Γ computations are cancelled and the gate needs to define them afresh.

2.1.6. *Accessible nodes.* Accessible nodes are defined at each stage very much as is the custom for tree constructions. For the $G_{\Phi,i,j}$ requirement, define

$$\ell(\Phi, i, j)[s] \;=\; \max\{\beta : \Phi(B \oplus A_i; \beta) \downarrow = \Phi(B \oplus A_j; \beta) \downarrow [s]\}.$$

We say that ρ is *expansionary* at s if it is accessible at s, and if for all $t < s$ at which ρ was accessible, $\ell(\Phi, i, j)[s] > \ell(\Phi, i, j)[t]$.

For future reference we note

LEMMA 6.6. *If ρ is expansionary at s then $\Gamma_\rho(B)[s]$ is doubly believable.*

PROOF. Unbelievable computations are removed, hence $\Gamma_\rho(B)[s]$ is believable. Let

$$m + 1 \;=\; \operatorname{dom} \Gamma_\rho(B)[s];$$

and say

$$\Phi(B \oplus A_i; b_m) \downarrow \; [s] \;=\; \Gamma_\rho(B, m) \, [s] \;=\; \sigma.$$

Now the computation $\Gamma_\rho(B, m) = \sigma$ was enumerated into Γ_ρ by ρ at an earlier stage $t < s$, at which ρ was accessible. Now $b_m[t] = b_m[s]$ (or the computation becomes unbelievable). ρ defines a Γ computation only if there is agreement between A_i and A_j; thus $\ell(\Phi, i, j)[t] \geqslant b_m[s]$. Thus $\ell(\Phi, i, j)[s] > b_m[s]$. This implies that $\Phi(B \oplus A_j; b_m) \downarrow = \sigma[s]$ as required. □

As usual we have

LEMMA 6.7. *If the hypothesis of $G_{\Phi,i,j}$ is correct then* $\lim_{s \to \alpha} \ell(\Phi, i, j)[s] = \alpha$.

PROOF. Let $\beta < \alpha$. We have $\Phi(B \oplus A_i; \beta) = \Phi(B \oplus A_j; \beta)$, and these computations hold from some stage s onwards; after s we always have $\ell(\Phi, i, j) \geqslant \beta$. □

At stage s, after some initial action by P, we determine which nodes are accessible by induction. If η is a hole which is accessible, then $\eta^\frown 0$ is accessible; if ρ is a gate and is expansionary at s, then $\rho^\frown 0$ is accessible, otherwise $\rho^\frown 1$ is. Note that the calculations of ℓ are done after P's action, and after some cascading took place. The definition of the accessible nodes ends when we get to a node which requires attention, or to a node that was just initialized (by P, or during cascading).

2.1.7. *The postal service comes to rescue.* Now here's a delicate point. We saw how to protect a follower's realization by finely managing its entourage's movements and cancelling weaker balls and markers. The idea is that at every stage from the follower's realization onward, all of the η-gates act in concert to protect x; we manage the entourage in such a way so that no η-gates would enumerate a troublesome marker, even if the computations making things believable grow and prosper.

There is a snag though, and it is the cascading. At any stage, to follow P's strategy, we dump many numbers into B. x can prevent B from directly injuring the computation realizing x by imposing restraint on P. However, P's action may also make some Γ_ρ computations (for ρ an η-gate) unbelievable, prompting markers γ_ρ to go into B, and perhaps destroy x's realization. If $\operatorname{dom}\Gamma_\rho(B)[s]$ is finite, then the total restraint that needs to be imposed on P to prevent such a chain of events will be finite (i.e. bounded below $b_\omega[s]$). But at the beginning of some limit stages it could be the case that $\operatorname{dom}\Gamma_\rho(B) = \omega$, and then it is possible that enumerating any $b_m[s]$ into B will prompt a cascading effect which will destroy x's realization. The amount of restraint needed to prevent this occurrence is infinite, and this is bothersome for P; in fact, this would derail the entire construction.

The solution is to protect, for every follower x and η-gate ρ, a computation $\Gamma_\rho \restriction n_\rho(x)$ (n_ρ is called a *stamp*). The ultimate goal is that together, all stamps for x set a barrier γ, making sure that for all η-gates ρ, $\Gamma_\rho \restriction n_\rho(x)$ is believed due to some computation with use less than γ, but all the markers for computations $\Gamma_\rho(n_\rho(x))$ are greater than γ. Thus all cascading must stop at $n_\rho(x)$.

The main idea in carrying out this strategy is that at every stage at which balls of x's entourage are enumerated into their sets, we *activate* the stamps $n_\rho(x)$, which simply means enumerating $\gamma_\rho(n_\rho(x))$ into B, thereby removing unprotected Γ_ρ computations. Thus at the end of the stage, we have, for all η-gates ρ, $\operatorname{dom}\Gamma_\rho(B) = n_\rho(x)\ [s]$. That is, we ensure that there are no dangerous markers because all computations are protected.

This action, though, is not enough, and we now explain why. Suppose that at stage s, balls from x's entourage enter their sets, and that the new end of the entourage lies at gate ρ^* (so ρ^* and all gates below it become safe at s). Now all $\Gamma_\rho \restriction n_\rho(x)$ are only singly-believable, since it is most likely that computations for one side were injured by the aforesaid enumeration.

These computations come back slowly and gradually, from the lower η-gates up to ρ^*. But by the time we get to ρ^* (i.e. at the first time t it is expansionary since s), lower η-gates ρ have defined computations $\Gamma_\rho(n_\rho(x))$ (for they cannot wait for ρ^* or for x), with uses $\gamma_\rho(n_\rho(x))$ which are smaller than the use of the new Φ computation for ρ^* (on the side that was injured). Now if we do nothing and activate all stamps at a later stage, we shoot ourselves in the foot, because we eliminate one side for ρ^*, while letting balls cross ρ^* which will eventually destroy the other side, leaving us "bald on both sides". We violated the principle of protecting both sides when ρ^* is safe, not by appointing small markers, but rather because of the necessary increase of the barrier γ. The solution is immediate: increase $n_\rho(x)$. We can do this because after s, all $\rho \subseteq \rho^*$ are safe. Note that for $\rho \supsetneq \rho^*$, we do not (and cannot) increase $n_\rho(x)$ since they are unsafe; but neither do we wish to, because they were not accessible between s and t, and so didn't appoint new markers. Also note that such gates ρ are not bothered by the entire pandemonium, since they are only protecting one side, which existed before s.

We can thus describe the complete instructions for handling the stamps. Let η be a hole and x be a follower. At stage w_0, x is realized and released. At this stage we issue the stamps: we set $n_\rho(x) = \operatorname{dom}\Gamma_\rho(B)$ for all η-gates ρ. We noted earlier that this is finite.

Now at every stage s at which balls from x's entourage get enumerated into their target sets, we activate all stamps (enumerate all $\gamma_\rho(n_\rho(x))$ into B for ρ such that $\Gamma_\rho(B, n_\rho(x)) \downarrow [s]$.)

At each such stage s, we let $\rho^* = \rho^*(x, s)$ be the gate at which the new end of x's entourage is waiting. At the first stage $t > s$ at which ρ^* is expansionary, we perform the *echo effect*: we redefine, for all $\rho \subseteq \rho^*$, $n_\rho(x) = \operatorname{dom}\Gamma_\rho(B)[t]$ [We note that $\Gamma_\rho(B)$ is doubly believable at t because ρ is expansionary at t]. We also initialize all nodes weaker than η and declare that ρ^* is the last node accessible at t.

Note that $n_\rho(x)$ is only updated once after each stage at which x enumerates balls, which we showed happens only finitely many times. Thus we have

LEMMA 6.8. *$n_\rho(x)$ is redefined finitely many times.*

This implies that at limit stages, $n_\rho(x)$ is well-defined and finite.

2.1.8. *Restraint.* Recall that the strategy to satisfy the P requirement is at every stage s, to enumerate all but finitely many numbers $< \omega s$ into B. This clearly suffices. However, it is easily noticed that such wild action may be harmful to other requirements, and so, most nodes will wish to impose restraint on P. We need to make sure, though, that such restraint is indeed finite.

Thus at the beginning of the stage we initialize all nodes at levels greater than $n(s)$. Also, for every hole η, we eliminate all but the strongest $n(s)$ followers.

The restraint on P at the beginning of s consists of the following:

(1) For every gate ρ, let $m_\rho[s] = \min\{n(s), \operatorname{dom}\Gamma_\rho(B)\,[s] - 1\}$. Suppose that the computation $\Gamma_\rho(B, m_\rho[s])$ was defined at some $t < s$. Then ρ restrains P from enumerating numbers below ωt into B.
(2) For every hole η and uncancelled follower x for η (even if already enumerated), η restrains P from enumerating numbers below ωt, where $t < s$ is the last stage at which x received attention.
(3) P restrains itself from enumerating $b_{n(s)}[s]$ into B.

The total restraint on P is denoted by $r(s)$. To act, P enumerates every number between $r(s)$ and ωs into B.

As defined, the restraint on P imposed by any ball or node is finite (in the sense that it protects only a finite initial segment of $\neg B$ from entering B. The instructions will show that at the beginning of every stage, after removing balls and initializing nodes, there are only finitely many balls on the machine and finitely many noninitialized nodes. We thus have:

LEMMA 6.9. *For every s, if $\neg B \cap \omega s$ is infinite, then $r(s) < \omega s$.*

2.1.9. *Attention.* Suppose that η is a hole which is accessible at stage s. We say that η *requires attention* if it is not yet satisfied, and if one of the following holds:

(1) η has a follower x on the machine, and x's entourage's end is waiting at a gate.
(2) η has a follower x waiting at the hole, and x is realized.
(3) There is no follower waiting at the hole.

We also say that a gate ρ requires attention if

4. ρ is an η-gate for some hole η, which has a follower x on the machine; at some stage $t < s$, numbers from x's entourage were enumerated into their sets, and s is the least stage after t at which ρ is expansionary, and the new end of x's entourage is now lying at ρ.

In cases (1) and (2) we say that x requires attention, and we give attention to the strongest such x (while removing all weaker followers). In case (1), we let the permissible tail of the entourage's end move to the next η-gate, or to the permitting bin if there is no such gate. In case (2), We let x's entire entourage move to the highest η-gate; we issue stamps.

In case (3), we appoint a large new follower x and place it at the hole. If ρ requires attention (case (4)), we perform the echo effect for x: we redefine stamps as described above, and initialize nodes weaker than η.

2.1.10. *Enumeration.* After P's action, and before defining accessible nodes, we check to see if some balls in the permitting bin are permitted. If there are some such, we choose the strongest follower x (from the strongest hole η) which has numbers permitted; we initialize nodes weaker than η, remove followers for η which are weaker than x, enumerate the numbers into their sets, and activate all stamps $n_\rho(x)$.

If a follower x was enumerated, then all other followers for its hole are cancelled, and the hole declares victory and considers itself satisfied (until the next stage at which it is initialized) [If x is ever cancelled, it is because the hole is initialized]. We will see that indeed if η declares victory and is not later initialized then its requirement is met.

2.1.11. *Defining Γ_ρ.* Suppose that ρ is expansionary at s. Then when we get to ρ, we wish to extend Γ_ρ. If $m = \operatorname{dom} \Gamma_\rho(B)[s]$ is under agreement (which means that $\Phi(B \oplus A_i; b_m) \downarrow = \Phi(B \oplus A_j; b_m) \downarrow [s]$), then we define $\Gamma_\rho(B, m)$ with the correct value and with a new, large marker.

2.2. Construction. We give the instructions for stage s of the construction.

- Act for P: initialize all nodes of length $\geqslant n(s)$. Also, remove all followers for holes η, except for the strongest $n(s)$ many followers. Calculate the restraint on P, and let P dump balls into B. Winnow entourages, to make them finite: remove all balls (in all entourages), which are at place

k in their entourage, where $b_k[s] > r(s)$. [Note that a follower is never winnowed]. Now remove unbelievable Γ_ρ computations and cascade.

- See if some balls in the permitting bin are permitted. If so, enumerate balls, initialize as described, and skip the next sub-stage. Remove unbelievable Γ_ρ computations and cascade.
- Find which nodes are accessible and require attention; extend Γ_ρ functionals for ρ which are expansionary (where it is possible) and then act for the first node which requires attention. Perform the echo effect if necessary.
- Appoint new traces at the end of every entourage.

That's the construction.

2.3. Verifications.

2.3.1. *Success of P.* At every stage, the restraint on P is finite, which implies that $\operatorname{otp} \neg B \cap s < \omega$ for all $s < \alpha$; thus $\operatorname{otp} \neg B \leqslant \omega$.

LEMMA 6.10. $\operatorname{otp} \neg B \geqslant \omega$.

PROOF. By induction on k, we show that $|\neg B| \geqslant k$. Suppose this holds for k. There is some stage $s_k > \beta_{k+1}$ such that for $i < k$, $b_i[s_k] = b_i$. Consider the stage $t_k > s_k$ at which $b_k[s_k]$ enters B (if there is no such stage then we're done). After P's action we have $b_k[t_k] = \omega t_k$; this number is never chosen as a marker.

After stage $t_k > \beta_{k+1}$, P is restrained from enumerating $b_k[t_k]$ into B. Thus $b_k[t_k] = b_k$ and $|\neg B| \geqslant k + 1$. □

We also note that the A_i are amenable; at stage s, after P's action, we have only finitely many balls on the machine. New balls are appointed large. Hence $A_i \restriction s$ differs from $A_i[s] \restriction s$ by only finitely many numbers, hence is α-finite.

The same observation gives us even more.

LEMMA 6.11. *For all X, if $A_i \leqslant_{w\alpha} X$ then $A_i \leqslant_\alpha X$.*

PROOF. To enumerate $A_i \restriction s$, X can first look at the construction at stage s. A look on the machine reveals finitely many balls which have a chance to enter A_i; by assumption, X can tell which ones will, and so determine $A_i \restriction s$. □

2.3.2. *The concert is sonorous and harmonious.*

LEMMA 6.12. *Suppose that x is a follower for η which works for $H_{\Psi,k}$, and that x is realized at some stage and then dropped from its hole. Then as long as x is not cancelled (even after it is enumerated), it is still realized by the original computation.*

PROOF. Suppose that x is dropped from the hole at t_0. Let $s_1, s_2, \dots, s_N$ be the stages at which members of x's entourage are enumerated into their target sets, and let t_i be the stage at which the echo effect (of s_i) occurs. We let, for $i < N$, ρ_i^* be the η-gate at which the end of x's entourage lies at the beginning of t_i (so ρ_0^* is the highest η-gate).

For every η-gate ρ (working for $G_{\Phi,i,j}$) and stage s at which there are numbers of x's entourage below ρ, there is one side $l_\rho(x)[s] \in \{i, j\}$ which is protected from injury: no balls targeted for $l_\rho(x)[s]$ crossed ρ at the last time balls from x's entourage crossed ρ.

At stage s, the computation $\Phi(B \oplus A_l; n_\rho(x))\,[s]$ is *protected* for $l = l_\rho(x)\,[s]$ or for both $l \in \{i, j\}$ if $s \in (t_i, s_{i+1}]$ and $\rho \subseteq \rho^*$.

A quick inspection shows that the only danger to protected computations may come from markers for η-gates. For by definition, balls from x's entourage cannot destroy a protected computation; the moment that balls from x's entourage, targeted for A_i, cross an η-gate ρ, The computation $\Phi(B \oplus A_i; n_\rho(x))$ ceases to be protected. Of course stronger balls don't move before x is cancelled; the same goes for marker for gates ρ such that $\rho^\frown 0 <_L \eta$. Balls weaker than x and markers weaker than η are always cancelled when x receives attention, in particular in every t_i and s_i, so are large. Also note that η imposes sufficient restraint on P to prevent this dumping requirement from intervening.

By induction on $s \geqslant t_0$ we can show that all protected computations are preserved (from the moment they are declared protected until they cease to be protected) and that all markers $\gamma_\rho(n_\rho(x))\,[s]$, if defined, are large. This is because after each t_i and s_i, for all η-gates ρ, we have $\Gamma_\rho(B, n_\rho(x)) \uparrow\ [s]$, so new markers are chosen large. Thus by induction we can show that if $\Phi(B \oplus A_l; n_\rho(x))\,[s]$ is protected, then $\Gamma_\rho(B; n_\rho(x))\,[s]$ is l-believable and so small markers are not enumerated into B. □

2.3.3. *Fairness, true path and success of the holes.*

LEMMA 6.13. *If η is a hole which is accessible α many times and eventually never initialized, then eventually η stops acting.*

PROOF. If there is a follower which is never realized or cancelled, then after appointing it, η never acts (and wins). If after $r^* = \mathtt{init}(\eta)[\alpha]$, a follower for η is enumerated into its target set (say at stage $s > r^*$), then at s, all other followers are removed and realization of the follower is preserved; after s, η does not act, and it wins.

Assume that no follower is ever enumerated into the target set after r^* and that every appointed follower is later either cancelled or realized (in which case some entourage part gets stuck in the permitting bin).

Let W be an r.e. set defined as follows: $s > r^*$ is enumerated into $W^{[m]}$ if an mth follower for η at s receives attention at s.

For all $t > \beta_m$ we have $n(t) > m$. Thus after β_m, no k^{th} follower for η, for $k \leqslant m$, can be removed by P. Every follower receives attention finitely many times; it follows (by induction on $k \leqslant m$) that there are only finitely many k^{th} followers appointed after stage $\max\{\beta_m, r^*\}$. It follows that $W^{[m]} \setminus \beta_m$ is finite. By the properties of f and g, there is some large m such that $g(m) > \max W^{[m]}$.

Let x be the last m^{th} follower appointed. We claim that x cannot have balls stuck indefinitely in the bin. For suppose there are, and that they got to the bin at stage s. s enters $W^{[m]}$, and so $g(m) > s$. Since $g(m)[s] < s$, we must have a change in $g(m)$ at some stage $t > s$, at which x is permitted and the balls would be enumerated into their sets; contradiction. □

We define the true path, as usual, to be the left-most nodes which are visited α many times. We can now prove fairness.

LEMMA 6.14. *A node on the true path is not injured α many times.*

PROOF. By induction on the length of the node. Suppose this holds for all $\tau \subsetneq \sigma$ where σ is on the true path. We know that eventually every $\tau \subsetneq \sigma$ stops initializing σ.

Of course, after some stage P does not injure σ.

Let s_0 be a stage after which neither P nor any $\tau \subsetneq \sigma$ injure σ, and after which no node to the left of σ is accessible. At s_0, there are finitely many nodes to the left of σ which are not initialized at s_0.

Holes $\eta <_L \sigma$ have finitely many followers at s_0, and do not appoint any new ones. Each such follower initializes σ at most twice (when elements of its entourage are enumerated into their sets, and later when an echo effect is performed for that enumeration).

Gates ρ such that $\rho^\frown 0 <_L \sigma$ have finitely many Γ_ρ computations defined at s_0, and never define more. ρ may injure σ each time it removes computations, but this occurs finitely many times.

Thus eventually no node to the left of σ injures σ. □

2.3.4. *D knows all.*

LEMMA 6.15. $A_i \leqslant_\alpha D$.

PROOF. Say that $x < \alpha$, and that D is asked whether $x \in A_i$ or not. First, D takes a look at stage x; if x is not a ball on the machine, targeted for A_i, and is not yet in A_i, then it will not enter A_i.

Suppose that x is a ball, which is part of the mth entourage for some hole η.

D can calculate a stage t after which m will never be permitted. After this stage, x will not be permitted by D. Thus if x is not yet in A_i at t and is still on the machine, D can guarantee that x will not be enumerated into A_i: it will either be cancelled, or parts of its entourage will for every lie in the permitting bin. □

We now need to see how D can compute B. P's action is not a problem here, as it is restrained by $n(s)$, and $\langle \beta_k \rangle$ is computable from D. Also, since D knows A_i, it can predict which balls will activate stamps. However, D cannot predict what kind of effect cascading may have on B. To overcome this problem we note that (as we needed for the success of the stamps), there are naturally occurring "dams" against cascading which protect B and can be detected by D. We define some γ to be *stable* at $s > \gamma$ if the following hold:

(1) For all ρ (working for $G_{\Phi,i,j}$), if $\Gamma_\rho(B,m)\downarrow [s]$ and $\gamma_\rho(m) < \gamma$, then there is an $l \in \{i,j\}$ such that $\Gamma_\rho(B,m)$ is l-believable at s by a Φ-computation with use less than γ.
(2) If x is an mth follower for its hole at s and there is a stamp $n_\rho(x)[s]$ such that $\gamma_\rho(n_\rho(x))[s] < \gamma$, then m will never be permitted after s.
(3) If $b_k[s] < \gamma$ then $s > \beta_k$.
(4) For $i < 3$, $A_i \restriction \gamma\, [s] = A_i \restriction \gamma$.

LEMMA 6.16. *If γ is stable at s then $B \restriction \gamma[s] = B \restriction \gamma$.*

PROOF. The third condition implies that after s, P will not enumerate numbers smaller than γ into B.

The second condition guarantees that no activations of stamps put numbers smaller than γ into B after s.

We show by induction on $t \geqslant s$ that no numbers smaller than γ enter B at stage t. Induction up to t, together with the fourth condition, imply that at t, all Γ_ρ computations with small markers (which necessarily were defined prior to s) are believable via the same computations which existed at stage s (by the first condition). It follows that at t, no such markers are put into B by gates removing unbelievable computations. □

LEMMA 6.17. *There are unboundedly many γ such that for some $t > \gamma$, γ is stable at t.*

PROOF. Suppose that at some stage w, a follower x^* for a hole η^* which lies on the true path is enumerated into its target set, and further suppose that η^* (and so x^*) is never injured after w. Let $\gamma = \omega w$.

Any $y < x^*$ which is on the machine at w is stronger than x^*, hence will never receive attention. All weaker balls are cancelled at w, and new ones appointed are large. It follows that for $i < 3$, $A_i \upharpoonright \gamma\,[w] = A_i \upharpoonright \gamma$.

At w, all stamps $n_\rho(x^*)$ for η^*-gates ρ are activated. As was shown in the proof of lemma 6.12, The computations $\Gamma_\rho \upharpoonright n_\rho(x^*)$ are permanent, so markers enumerated by η^*-gates after s are picked after w hence are greater than γ. No other gates may enumerate markers which are smaller than γ, and P too is prevented (by x^*) from enumerating numbers smaller than γ into B. It follows that $B \upharpoonright \gamma\,[w] = B \upharpoonright \gamma$.

Suppose that $s > w$ and suppose that ρ is a gate and that $\Gamma_\rho(B, m) \downarrow [s]$ with use $\gamma_\rho(m)[s] < \gamma$. Necessarily, this computation was defined before w and was believable at w, by a Φ-computation with use $< \gamma$ (at w, all computations have use $< \gamma$). This Φ-computation is preserved.

We can thus let s be a stage greater than w and β_k (where $\gamma = b_k[w]$) and late enough such that $n(w)$ will not be permitted after s. We claim that γ is stable at s. This suffices since we can clearly find such γ as large as we like.

We know that the first, third and fourth conditions hold. We show that the second condition holds too. Suppose that at s, x is a follower on the machine and there is a stamp $n_\rho(x)[s]$ such that $\gamma_\rho(n_\rho(x))[s] < \gamma$. Necessarily, the computation $\Gamma_\rho(B, n_\rho(x)[s])$ is defined prior to stage w, and does not change until s. Every follower y appointed at $t > w$ which defines a stamp $n_\rho(y)$ will have $n_\rho(y) = \operatorname{dom}\Gamma_\rho(B)[t] > n_\rho(x)[s]$. Hence x is already on the machine at w, so is the mth follower for its hole, $m < n(w)$ (other followers are removed at w). After s, x will not be permitted. □

Finally, we can see that D can tell whether some γ is stable at a stage s. The first condition is computable. The second can be decided by D because it controls the permitting. The third, because D can compute β_n; and the last, because $A_i \leqslant_\alpha D$. We have proven

LEMMA 6.18. *$B \leqslant_\alpha D$.*

2.3.5. *The Top.*

LEMMA 6.19. *$A_i \leqslant_\alpha B \oplus A_j \oplus A_k$.*

PROOF. To find if $x \in A_i$, we first look at stage x, to see if x is a ball on the machine, targeted for A_i. If so, at that stage, x is the nth element of its entourage. Using B we can find a stage $s > x$ such that $b_{n+1}[s] = b_{n+1}$. If x is still on the machine at s, then a trace for x will never be winnowed after x.

The usual argument shows that after s, x has only finitely many traces, and the last one either enters its set at the same time x enters A_i, or doesn't, in which case $x \notin A_i$. □

2.3.6. *Success of the Gates.*

LEMMA 6.20. *Suppose that ρ working for $G_{\Phi,i,j}$ is on the true path, and assume that the hypothesis of $G_{\Phi,i,j}$ holds. Then $\Gamma_\rho(B) = \Phi(B \oplus A_i)$.*

PROOF. ρ is expansionary α many times.

By induction on n, we show that $\Gamma_\rho(B, n) \downarrow = \Phi(B \oplus A_i; b_n)$. Assume up to $n-1$ and assume that after s_0, ρ is never initialized; for $m < n$, $\Gamma_\rho(B, m)$ computations are defined with correct markers (so they will not end up in B) and correct Φ-computations from both sides; $b_n[s_0] = b_n$, and $\Phi(B \oplus A_l; b_n)$ is correct for both $l \in \{i, j\}$. Assume also that s_0 is expansionary.

At every expansionary stage $s \geqslant s_0$, $\Gamma_\rho(B; n)$, if not already defined, gets defined (because the agreement between $\Phi(B \oplus A_i)$ and $\Phi(B \oplus A_j)$ is permanent). At s_0, if we have an incorrect computation, then it gets removed; all later computations are always doubly believable (and give the correct answer). Such a computation cannot be removed by P (it is prohibited; $n(s) > n$). The believability also implies that such a computation cannot be cancelled during cascading. The only reason that $\Gamma_\rho(n)$ may be cancelled again is that $n = n_\rho(x)$ for some follower x.

At s_0, there are finitely many xs on the machine such that $n_\rho(x) = n$, thus eventually (after some s_1), none of these balls activate their stamps.

If at $s \geqslant s_0$, some follower y defines $n_\rho(y)$, then ρ is expansionary at s. Thus before the definition, ρ got to define $\Gamma_\rho(n)$. Thus $n_\rho(y) > n$. Thus after s_1, no stamp $n_\rho(x) = n$ will be activated.

Thus we eventually get a permanent computation. □

APPENDIX A

Basics

It is convenient to use Jensen's J_α hierarchy. This is a cumulative hierarchy of the constructible universe L. Readers who are unfamiliar with this hierarchy, may ignore the letter J throughout this work and replace it by the letter L. A formal definition of the J_α hierarchy and an investigation of its basic properties can be found in texts such as Jech [**Jec03**] or Dodd [**Dod82**]. The advantage of using J_α over L_α is that it behaves nicely with respect to the notion of the projectum, which is discussed in appendix C. Essentially, $J_{\alpha+1}$ is obtained from J_α by closing $J_\alpha \cup \{J_\alpha\}$ under a finite collection of *rudimentary* functions (similar to the Gödel operations). $J_\alpha \cap \mathrm{On} = \omega\alpha$, the αth limit ordinal.

For our purposes, let an *amenable structure* be a structure of the form $(J_\beta, \in, X)$ where $\beta \geqslant 1$ is an ordinal and $X \subset J_\beta$ is *amenable* (set) for J_β, that is, for every $x \in J_\beta$, $X \cap x \in J_\beta$. Amenable sets are also called *regular*.

Since $\in$ is a relation in all amenable structures, we suppress it in notation and refer to the structure (J_β, X). We write J_β for both the set and the structure $(J_\beta, 0)$. If $M = (J_\beta, X)$ is an amenable structure then we sometimes use M to denote the universe J_β.

The *Levy hierarchy* for formulas in the language of set theory, augmented by a unary predicate, is the familiar one. For any amenable structure M, $\Delta_0(M)$ denotes the collection of subsets of M defined by bounded formulas, with parameters in M. Similarly we have $\Sigma_n(M)$, $\Pi_n(M)$.

FACT. If M is an amenable structure then every $A \in \Delta_0(M)$ is amenable for M.

Various set-theoretic notions have also recursion-theoretic names. Let M be an amenable structure. A set x is *M-finite* if $x \in M$ (we usually say β-finite instead of J_β-finite). The sets which are $\Sigma_1(M)$ are also called *M-recursively enumerable*. $\Delta_1(M)$ sets (i.e. those which are both M-r.e. and M-co-r.e.) are called *M-recursive*. A partial function $f : M \to M$ is M-recursive if it is M-r.e. (as a set of pairs).

FACT. For every β, there is a β-recursive bijection between J_β and $\omega\beta = J_\beta \cap \mathrm{On}$.

Thus, for the purposes of recursion theory, we do not care if the sets considered are subsets of J_β or of $\omega\beta$.

We write β-r.e. instead of J_β-r.e. and β-recursive instead of J_β-recursive. When β is fixed we omit the prefix β.

We never write "finite" instead of β-finite. It is in fact the theme of this work to exhibit the fundamental difference between finite and β-finite.

An amenable structure M is *admissible* if the image of every M-finite set, under an M-recursive function, is bounded (say in $<_J$). An amenable set $X \subset J_\beta$ is called *admissible* (when β is fixed) if (J_β, X) is admissible. Admissible sets are also called "regular and hyperregular". An ordinal α is admissible if J_α is an admissible structure. Throughout this thesis, α always denotes an admissible ordinal.

FACT. Let $M = (J_\alpha, X)$ be admissible.

(1) $\alpha = \omega\alpha$; $J_\alpha = L_\alpha$.
(2) M satisfies Δ_1-comprehension: every $A \in \Delta_1(M)$ is amenable for M.
(3) M satisfies Σ_1-bounding: the image of an M-finite set under an M-recursive function is M-finite. In fact, if f is partial M-recursive, x is M-finite and $x \subset \operatorname{dom} f$, then $f \upharpoonright x$ is M-finite.
(4) If A is M-r.e. then $\forall x \in y\, A$ is also M-r.e.
(5) All familiar elementary theorems of ω-recursive function theory hold (enumeration theorem, s-m-n theorem, definition by induction, recursion theorem, etc.)

A *semi-admissible* structure is an amenable structure (J_α, X) where α is admissible. We will almost always work with semi-admissible structures.

Functionals and Reducibilities. Let $M = (J_\beta, X)$ be an amenable structure. An *M-enumeration functional* is an M-r.e. set whose elements are pairs (p, x), where p is a *partial string*, that is, a function with domain a set of ordinals and range 2.

If Ψ is an M-enumeration functional and p is any partial string (not necessarily M-finite), we let

$$\Psi(p) \;=\; \{x : \exists p_0 \subset p\; [(p_0, x) \in \Psi]\}.$$

Let $A \subset \omega\beta$. We say that $B \subset M$ is *M-r.e. in A* if there is some M-enumeration functional Ψ such that $B = \Psi(A)$. A partial function is *M-weakly recursive* in A if it is M-r.e. in A as a set of pairs. A set $B \subset \alpha$ is M-weakly recursive in A (we write $B \leqslant_{wM} A$) if it is both M-r.e. and M-co-r.e. in A.

An *M-functional* (also known as a *Turing functional* or as a *strong functional*) is an M-enumeration functional whose range also consists of partial strings. When we use an M-functional Ψ we use $\cup\Psi(p)$ instead of $\Psi(p)$ (and write $\Psi(p)$ for the former, just to confuse things. However we only use this notation if $\Psi(p)$ is a function.)

Let $A, B \subset \omega\beta$. B is *strongly M-r.e.* (or *tamely* M-r.e.) in A if

$$\{K \subset B : K \in M\}$$

is M-r.e. in A (Here we regard B as a *set* rather than as its characteristic function).

B is *M-recursive* in A (we write $B \leqslant_M A$) if both B and $\neg B = \omega\beta \setminus B$ are strongly M-r.e. in A. B is M-recursive in A iff

$$\{p \in M : p \text{ is a partial string } \;\&\; p \subset B\}$$

is M-r.e. in A.

The relation $\leqslant_M$ is a reflexive and transitive relation on the subsets of α. The equivalence classes are called M-degrees, or X-degrees (when α is understood). J_α-degrees are usually called α-degrees.

LEMMA A.1. *Let $M = (J_\beta, X)$ be amenable and let $A \subset \omega\beta$ be amenable. Let $B \subset M$. Then B is M-r.e. in A iff it is (M, A)-r.e.*

Proof. Suppose that $B = \Phi(A)$ where Φ is an M-enumeration functional. Then $x \in B \iff \exists p \subset A\ [(p,x) \in \Phi]$ which is $\Sigma_1(M,A)$.

Suppose that $B \in \Sigma_1(M,A)$ (defined by the formula $B(x)$; say B is over S_γ, $\gamma < \omega\beta$.) Let

$$\Phi = \{(p,x) : \operatorname{dom} p > \gamma \ \&\ (S_{\operatorname{dom} p}, X \upharpoonright \operatorname{dom} p, A \upharpoonright \operatorname{dom} p) \models B(x)\}.$$

Then $\Phi(A) = B$. □

Thus $B \leqslant_{wM} A$ iff B is (M,A)-recursive.

Lemma A.2. *Let $M = (J_\beta, X)$ be amenable and let $A \subset \omega\beta$ be amenable. For all $A, B, C \subset \omega\beta$, B is M-r.e. in $A \oplus C$ iff it is (M,A)-r.e. in C.*

Note that this follows from lemma A.1 if C is amenable.

Proof. Suppose that Φ is an M-enumeration functional such that $A = \Phi(B \oplus C)$. Let

$$\Psi = \{(p,x) : \exists q \subset A\ [((q,p),x) \in \Phi]\}.$$

Ψ is (M,A)-r.e. and $\Psi(C) = B$.

Suppose that Ψ is (M,A)-r.e. and that $\Psi(C) = B$. Then by lemma A.1, Ψ is M-r.e. in A; say Θ is M-r.e. and $\Theta(A) = \Psi$. Let

$$\Phi = \{((q,p),x) \in M : (q,(p,x)) \in \Theta\}.$$

Then Φ is M-r.e. and $\Phi(A \oplus C) = B$. □

Corollary A.3. *Let $M = (J_\alpha, X)$ be amenable and let $A \subset \alpha$ be amenable. For all $B, C \subset \alpha$, $B \leqslant_M A \oplus C$ iff $B \leqslant_{M,A} C$ and $B \leqslant_{wM} A \oplus C$ iff $B \leqslant_{w(M,A)} C$.*

Thus the (M,A)-degrees are isomorphic to the cone of M-degrees above A.

Lemma A.4. *If M is admissible, then every M-r.e. set is also strongly M-r.e.* □

Thus if (J_α, A) is admissible then for all B, $B \leqslant_{w\alpha} A$ iff $B \leqslant_\alpha A$.

Regular Functionals. From now, we fix an admissible ordinal α and work, for simplicity, in J_α. All results can be relativized to admissible structures (J_α, X).

An enumeration functional Ψ is called *regular* if its domain consists of strings – partial strings whose domain is an ordinal.

Lemma A.5. *Let Ψ be an enumeration functional. There is an regular enumeration functional Φ such that for all amenable $A \subset \alpha$, $\Phi(A) = \Psi(A)$.*

An r.e. index for Φ can be uniformly obtained from an r.e. index for Ψ.

Proof. Let $\mathcal{S}$ be the set of α-finite strings, and let

$$\Phi = \{(p,x) : p \in \mathcal{S} \ \&\ x \in \Psi(p)\}.$$

Let p be a string. $\Phi(p) \subset \Psi(p)$. On the other hand, if p_0 is a partial string, $x \in \Psi(p_0)$ and p is any string extending p_0 then $x \in \Phi(p)$. The conclusion follows. □

Claim A.6. *Let $A, B \subset \alpha$ and suppose that B is amenable. Then $B \leqslant_\alpha A$ iff*

$$\{B \upharpoonright \beta : \beta < \alpha\}$$

is r.e. in A.

PROOF. Easy, because given a string p we can (uniformly) enumerate all partial strings extended by p. □

A functional is regular if it is regular as an enumeration functional and if its range consists of strings.

LEMMA A.7. *Let $A, B \subset \alpha$ be amenable. Then $B \leqslant_\alpha A$ iff there is some regular functional Φ such that $B = \Phi(A)$.*

An index for Φ and an index for a reduction of B to A are uniformly interchangeable.

PROOF. If Ψ is a regular enumeration functional which enumerates, with oracle A, all α-finite partial strings extended by B, then simply take $\Phi = \Psi \cap \mathcal{S}^2$, where as before $\mathcal{S}$ is the set of α-finite strings. On the other hand, if Φ is a regular functional and $\Phi(A) = B$ then considered as an enumeration functional, $\Phi(A)$ enumerates $B \upharpoonright \beta$ for cofinally many βs; one can now enumerate all strings extended by strings in $\Phi(A)$. □

Recursively Enumerating Recursively Enumerable Sets. Suppose that A is an r.e. set. A is identified with the formula defining it over J_α. Assume the parameters used lie in J_β ($\beta < \alpha$). β can be uniformly obtained from an index for A (which can be taken to be the formula A. In this way we identify the set A with its index).

We let $A[s] = 0$ for $s < \beta$ and for $s \geqslant \beta$ we let

$$A[s] = \{x \in J_s : J_s \models A(x)\}.$$

The sequence $\langle A[s] \rangle_{s<\alpha}$ is increasing, continuous, recursive and

$$\bigcup_{s<\alpha} A[s] = A.$$

Nice Functionals. If Φ is an enumeration functional then for all s, $\Phi[s]$ is an enumeration functional too. We write $\Phi(p)\,[s]$ for $\Phi[s](p)$. The relation $x \in \Phi(p)\,[s]$ is recursive.

DEFINITION. A regular enumeration functional Φ is *nice* if for all α-finite p and x, $(p, x) \in \Phi$ iff $x \in \Phi(p)[\operatorname{dom} p]$.

If Φ is a nice enumeration functional then for all α-finite p and x, $x \in \Phi(p)$ iff $(p, x) \in \Phi$; this is because $\Phi[s]$ is increasing. If Φ is nice then Φ, and so the relation $x \in \Phi(p)$, are recursive.

LEMMA A.8. *Let Ψ be a regular enumeration functional. There is some nice enumeration functional Φ such that for all amenable $A \subset \alpha$, $\Phi(A) = \Psi(A)$.*

A recursive index for Φ can be uniformly obtained from an r.e. index for Ψ.

PROOF. Let

$$\Phi = \{(p, x) : x \in \Psi(p)\,[\operatorname{dom} p]\,\}.$$ □

Note that if Φ is nice then the map $p \to \Phi(p)$ (ranging over α-finite strings p) takes values in J_α and is recursive.

DEFINITION. A functional Φ is *nice* if it is nice, considered as an enumeration functional, and furthermore:

(1) It is consistent: for all strings p, $\Phi(p)$ is a string.
(2) It is downward closed: For all strings p and q, if $q \subset \Phi(p)$ then $(p, q) \in \Phi$.

LEMMA A.9. *Let Ψ be a functional. There is some nice functional Φ such that for all amenable A, if $\Psi(A)$ is a function then $\Phi(A) = \Psi(A)$.*

PROOF. We may assume that Ψ is a nice enumeration functional. Now let

$$\Phi = \{(p,q) : \Phi(p) \text{ is a function } \ \& \ q \subset \Phi(p)\}.$$

Note that if $\gamma < \alpha$, $q \in J_\gamma$ and $q_0 \subset q$ is a string then $q_0 \in J_\gamma$. This preserves the fact that Φ is nice as an enumeration functional. □

As was discussed before, we have:

COROLLARY A.10. *Let $A, B \subset \alpha$ be amenable. Then $B \leqslant_\alpha A$ iff there is some nice functional Φ such that $\Phi(A) = B$.*

Sometimes an enumeration functional is used to enumerate graphs of functions. We then call it a *weak Turing functional.* Such a functional Φ is called nice if it is nice as an enumeration functional and it satisfies consistency: for all strings p, $\Phi(p)$ is a function (but not necessarily with ordinal domain). Similarly to lemma A.9, we get

LEMMA A.11. *Let Ψ be an enumeration functional. There is some nice weak functional Φ such that for all amenable A, if $\Phi(A)$ is a function then $\Phi(A) = \Psi(A)$.*

PROOF. Again we may assume that Ψ is a nice enumeration functional, and we let

$$\Phi = \{(p,x) : x \in \Phi(p) \ \& \ \Phi(p) \text{ is a function}\}.$$ □

Thus for amenable $A \subset \alpha$ and any $B \subset \alpha$, $B \leqslant_{w\alpha} A$ iff $\Phi(A) = B$ for some nice weak functional Φ.

Use. If Φ is a nice functional, p is a string and $\beta \leqslant \operatorname{dom}\Phi(p)$, then we let $\Phi(p;\beta) = \Phi(p) \upharpoonright \beta$ and $\phi(p;\beta)$ be the shortest $q \subset p$ such that $\beta \leqslant \operatorname{dom}\Phi(q)$. This is the *use* of the computation $\Phi(p;\beta)$. We sometimes confuse $\phi(p;\beta)$ with its domain. We write $\Phi(p;\beta)\downarrow$ if $\beta \leqslant \operatorname{dom}\Phi(p)$.

We have: if $\gamma < \delta$ and $\Phi(p;\delta)\downarrow$ then $\phi(p;\gamma) \leqslant \phi(p;\delta)$. If δ is a limit ordinal then $\phi(p;\delta) = \sup_{\gamma<\delta}\phi(p;\gamma)$. For all γ such that $\Phi(p;\gamma)\downarrow$ we have $\phi(p;\gamma) \geqslant \gamma$.

For particular values of a computed function we use the notation $\Phi(p,\gamma) = \Phi(p)(\gamma)$; here Φ can be either a strong or a weak functional. For a weak functional Φ we write $\Phi(p,\gamma)\downarrow$ if $\gamma \in \operatorname{dom}\Phi(p)$. We also define the use in an analogous way: $\phi(p,\gamma)$ is the least string $q \subset p$ such that $\Phi(q,\gamma)\downarrow$.

Recursive Approximation. Let A be an r.e. set and let $A[s]$ be its canonical recursive enumeration. When we take $A[s]$ as an oracle we regard it as a string of length s, and let $\Phi(A)[s] = \Phi(A[s])$ (note that this conflicts with our previous definition of $\Phi(p)[s]$, where the object enumerated was Φ rather than the oracle. Since we will use nice functionals, which are recursive, we will have no use for the original notation.)

Assume that A is amenable. By admissibility of α, we have that for every $\beta < \alpha$, $A \upharpoonright \beta \subset A[t]$ from some stage $s < \alpha$ onwards. Thus if $x \in \Phi(A)$ then from some s onwards, $x \in \Phi(A)[t]$. However it is of course possible that for some x and

s we have $x \in \Phi(A)[s]$ but $x \notin \Phi(A)$; x is enumerated at stage s by a computation that does not agree with A, which is later discarded. In fact if A is not low then this may occur for some x at unboundedly many stages s.

Miscellany.

THEOREM A.12 (Sacks). *If* $\mathbf{a}$ *is an r.e. degree then there is some amenable, r.e.* $A \in \mathbf{a}$.

PROOF. See [**Sac90**]. □

LEMMA A.13. *If* $A \leqslant_\alpha B$ *and* C *is r.e. in* B *then* C *is r.e. in* A. □

As in classical recursion theory, for $A, B \subset \alpha$ we say that $B \leqslant_{m\alpha} A$ if there is some recursive function f such that $f^{-1}B = A$.

LEMMA A.14. *Suppose that* B *is r.e. in* A *and that* $C \leqslant_{m\alpha} B$. *Then* C *is r.e. in* A.

PROOF. Suppose that f is recursive and that $f^{-1}B = C$. If $B = \Phi(A)$ then $C = \Psi(A)$ where

$$\Psi \;=\; \{(p, x) \,:\, (p, f(x)) \in \Phi\}.$$ □

Recursive Cofinality. Let Γ be a class of functions. We let

$$\mathrm{cf}_\Gamma(\alpha) \;=\; \min\{\gamma \,:\, \exists f \in \Gamma \;\; (f : \gamma \to \alpha \text{ is cofinal})\,\}.$$

For example, Γ may consist of the functions which are $\Sigma_n(J_\alpha)$ or more generally, $\Sigma_n(J_\alpha, A)$ for some amenable $A \subset \alpha$. α being admissible means $\mathrm{cf}_{\Sigma_1(J_\alpha)}(\alpha) = \alpha$.

Even if A is not amenable then we can let Γ be the class of all functions which are weakly recursive in A. In this case we write $\mathrm{rcf}(A)$ instead of $\mathrm{cf}_\Gamma(\alpha)$. Note that $\mathrm{rcf}(0') = \mathrm{cf}_{\Sigma_2(J_\alpha)}(\alpha)$.

APPENDIX B

The Jump

In what follows we confuse functionals and r.e. sets with their indexes. For any amenable A we let

$$A' = \{(\Phi, x) : x \in \Phi(A)\}$$

where Φ ranges over an effective list of nice enumeration functionals. A' is r.e. in A; in fact there is a nice enumeration functional $\Phi_{\mathtt{jump}}$ such that for all amenable A, $\Phi_{\mathtt{jump}}(A) = A'$. The usual diagonal argument shows that $A' \not\leqslant_{w\alpha} A$.

Characterizations of the Jump.

LEMMA B.1. *Let A be amenable. If $f \leqslant_{w\alpha} A$ and $B = \lim_{s\to\alpha} f$ then $B \leqslant_{w\alpha} A'$. If A is admissible and $B \leqslant_{w\alpha} A'$ then for some $f \leqslant_{w\alpha} A$, $B = \lim_{s\to\alpha} f$.*

PROOF. Suppose that $f \leqslant_{w\alpha} A$, $B = \lim f$. Let Ξ be a nice weak functional such that $\Xi(A) = f$.

$$W = \{(x, s) : \exists t > s\ [f(x,t) \neq f(x,s)]\}$$

is r.e. in A (since it is $\Sigma_1(L_\alpha, A)$ and A is amenable), say $W = \Psi(A) = (A')^{[\Psi]}$. Also say $(A')^{[\Lambda]} = A$. Let

$$\Phi = \left\{ (p, x) : \begin{array}{l} p \text{ is a partial string} \quad \& \\ \exists s\ [p(\Psi, (x,s)) = \mathtt{no} \ \ \& \ \ \Xi(p^{[\Lambda]}, (x,s)) = \mathtt{yes}] \end{array} \right\}.$$

Then $\Phi(A') = B$, so B is r.e. in A'. Similarly we have $\neg B$ r.e. in A' so $B \leqslant_{w\alpha} A'$.

Suppose now that A is admissible and that $B \leqslant_{w\alpha} A'$. By Sacks's theorem relativized to A we get an r.e. $C \equiv_\alpha A'$ which is amenable. Let Ψ be a nice weak functional such that $\Psi(C) = B$. Let $\langle C[s] \rangle$ be an enumeration of C, recursive in A. Let $f(x,s) = \Psi(C[s], x)$ if the latter converges and $\mathtt{no}$ otherwise. By the admissibility of A, for all $\beta < \alpha$, eventually $C[s] \restriction \beta \subset C$ so indeed $\lim f = B$. □

LEMMA B.2. *Assume that A is admissible. Then $B \subset \alpha$ is r.e. in A' iff it is $\Sigma_2(A)$.*

PROOF. If $B \in \Sigma_2(A)$, say $x \in B \iff \exists y\ C(x,y)$ where $C \in \Pi_1(A)$; $C(x,y) \iff (C, x, y) \notin A'$. It is now easy to see how to enumerate B with oracle A'.

Let B be r.e. in A'. Again by Sacks's theorem we get an r.e. $C \equiv_\alpha A'$ which is amenable; $B = \Phi(C)$ for some nice enumeration functional Φ. Let Ψ be a nice enumeration functional such that $C = \Psi(A)$. By admissibility of A we have that every α-finite $K \subset C$ is enumerated by a finite stage, so we have

$$x \in B \Leftrightarrow \exists K, L, p \left[\begin{array}{l} L \cap C = 0 \ \ \& \ \ p \subset A \ \ \& \ \ \forall x \in K\, ((p,x) \in \Psi \ \ \& \\ (L \times \{0\} \cup K \times \{1\}, x) \in \Phi \end{array} \right].$$

$L \cap C = 0$ is $\Pi_1(A)$ and all other clauses in the matrix are recursive in A. □

Lowness. A set A is *weakly low* if $A' \leqslant_{w\alpha} 0'$. Suppose that A is r.e. and amenable, and fix an effective enumeration $\langle A[s] \rangle$ of A. A nice enumeration functional Ψ is *weakly low* for A if $\Psi(A) = A'$ and in addition, for all x, if

$$\{s < \alpha \,:\, x \in \Psi(A)[s]\}$$

is unbounded in α, then $x \in A'$.

LEMMA B.3. *An amenable r.e. set A is weakly low iff there is some functional which is weakly low for A. .*

PROOF. Suppose that Ψ is weakly low for A. Let

$$W \;=\; \{(x, s) \,:\, \exists t > s \,:\, x \in \Psi(A)[t]\}.$$

$\neg A'$ is r.e. in W as $x \notin A'$ iff for some s, $(x, s) \notin W$. W is r.e., so $W \leqslant_\alpha 0'$. Thus $\neg A'$ is r.e. in $0'$. $A \leqslant_\alpha 0'$ thus A' is r.e. in $0'$; so $A' \leqslant_{w\alpha} 0'$.

Now suppose that A is weakly low; let $f(x)[s]$ be a recursive function whose limit is A'. Let

$$\Psi \;=\; \{(p, x) \,:\, (p, x) \in \Phi_{\texttt{jump}} \;\; \& \;\; f(x)[\operatorname{dom} p] = \texttt{yes}\}.$$

$\Psi(A) \subset A'$ and if $x \in A'$ and $s > \phi_{\texttt{jump}}(A, x)$ is such that $f(x)[s] = \texttt{yes}$ then $(A \upharpoonright s, x) \in \Psi$, thus $\Psi(A) = A'$.

Suppose that $x \notin A'$; say that for all $s > s_0$, $f(x)[s] = \texttt{no}$. Suppose that $A[s_1] \upharpoonright s_0 = A \upharpoonright s_0$. Then for $s > s_1, s_0$ we cannot have $x \in \Psi(A)[s]$; If $(p, x) \in \Psi$ and $p \subset A[s]$ then since $p \not\subset A$ we must have $\operatorname{dom} p > s_0$ and so $f(x)[\operatorname{dom} p] = \texttt{no}$, contradiction. □

We say that a set A is *low* if $A' \leqslant_\alpha 0'$.

LEMMA B.4. *For all A,*

$$\{K \in J_\alpha \,:\, K \subset \neg A'\}$$

is many-one reducible to $\neg A'$.

Thus for all B, if $\neg A'$ is r.e. in B then it is also strongly r.e. in B.

PROOF. The reduction is the (recursive) function f such that for all K, $f(K)$ is the functional which performs all computations $\Phi(A, x)$ for $(\Phi, x) \in K$ simultaneously and halts if one of these computations halt. Namely:

$$f(K) \;=\; \{p \,:\, \exists (\Phi, x) \in K \; [x \in \Phi(p)]\}. \qquad \square$$

COROLLARY B.5. *Suppose that A is admissible and that $B \geqslant_\alpha A$. Then $A' \leqslant_\alpha B$ iff $A' \leqslant_{w\alpha} B$.*

PROOF. A' is r.e. in A, thus strongly r.e. in A (lemma A.4), thus strongly r.e. in B. $\neg A'$ is r.e. in B hence strongly r.e. in B (lemma B.4). □

Again suppose that A is r.e. and amenable, and fix an effective enumeration $\langle A[s] \rangle$ of A. A nice enumeration functional Ψ is *low* for A if it is weakly low for A, and further, for all α-finite $K \subset A'$, eventually $K \subset \Psi(A)[s]$.

LEMMA B.6. *If A is r.e. and amenable and there is some functional which is low for A then A is low.*

PROOF. Suppose that Ψ is low for A. Then we already know that $A' \leqslant_{w\alpha} 0'$, so again we know that $\neg A'$ is strongly r.e. in $0'$; we need to show that A' is strongly r.e. in $0'$. But it is strongly r.e. in

$$W = \{(K, s) : \exists t > s : K \not\subset \Psi(A)[t]\}$$

which is r.e. □

LEMMA B.7. *If A is low, r.e. and admissible, then any functional which is weakly low for A is low for A.* □

The circle is closed via

LEMMA B.8. *If A is a low r.e. set then A is admissible.*

PROOF. By Shore [**Sho76b**], if $\mathrm{rcf}(A) < \alpha$ then $0'' \leqslant_\alpha A'$, so $A' \leqslant_\alpha 0'$ is impossible. Also, if A is r.e. and $\mathrm{rcf}(A) = \alpha$ then A is amenable so A is admissible. □

Thus an amenable r.e. set A is low iff some functional is low for A.

Uniform Lowness. A sequence of sets $\langle A_i \rangle_{i \in I}$ is *uniformly low* if there is a recursive sequence $\langle \Phi_i \rangle_{i \in I}$ such that for all i, $\Phi_i(0') = A_i'$. Any finite collection of low sets is uniformly low. If $\langle A_i \rangle$ are uniformly low and Ψ is an enumeration functional then

$$\{(i, x) : x \in \Psi(A_i)\}$$

is weakly recursive in $0'$.

APPENDIX C

The Projectum

Let $M = (J_\beta, X)$ be amenable. We define ϱ_M (also denoted $\varrho_{\beta,X}$) to be the least ordinal γ such that

$$\Sigma_1(M) \cap \mathcal{P}(\omega\gamma) \not\subset J_\beta.$$

ϱ_M is the Σ_1-*projectum* of M. We write ϱ_β for $\varrho_{\beta,0}$. ϱ_β is also denoted β^*.

CLAIM C.1. *If* $1 < \varrho_M < \beta$ *then* $\varrho_M = \omega\varrho_M$.

PROOF. If $\varrho_M < \beta$ then $M \models$ "$\omega\varrho_M$ is a cardinal", and so is closed under the function $\gamma \to \omega\gamma$. $\square$

If $\varrho_M = 1$ then we often confuse 1 and ω in this context (so we may write $\varrho_M = \omega$, meaning the same thing).

REMARK C.2. For all M, ϱ_M is admissible.

DEFINITION. We say that M is *1-sound* if there is some partial M-recursive function $f : \varrho_M \twoheadrightarrow M$.

LEMMA C.3 (Jensen). *For every* β, J_β *is 1-sound.*

(This is easy if β is admissible).

PROOF. Let p be the least (in $<_L$, or in any canonical well-ordering of finite sets of ordinals) finite set of ordinals such that some $A \subset \varrho_\beta$ is $\Sigma_1(J_\beta)$ definable with parameters from $\varrho_\beta \cup p$ but is not an element of J_β. Let h be the canonical Σ_1-Skolem function computed in J_β. Let $X = h``\varrho_\beta \cup \{p\}$, and let $\pi : X \to J_\gamma$ be the collapse. $A = \pi``A$ is $\Sigma_1(J_\gamma)$ so $\gamma = \beta$ (or else $A \in J_\beta$). Also, it is Σ_1-definable in J_β from $\varrho_\beta \cup \pi(p)$ so by minimality $\pi(p) = p$. It follows that $\pi = \text{id}$. So $h(-, p)$ is the desired function. $\square$

Suppose that M is 1-sound, and fix a canonical $f : \varrho_M \twoheadrightarrow M$, say the least one (in $<_L$). We let $A_M = f^{-1}B$ where B is the canonical $\Sigma_1(M)$-complete set. We let $M' = (J_{\varrho_M}, A_M)$.

If in turn M' is amenable and 1-sound then we say that M is 2-sound and we can proceed to define M'', etc.

LEMMA C.4. *(see* **[Cho84]**,**[Jen72]** *or* **[Dod82]***) For all* $\beta > 0$ *and* $n < \omega$, J_β *is* n*-sound.*

The idea is to imitate the proof of 1-soundness above, lifting $\pi : J_\gamma \to (J_\beta)^{(n-1)}$ to a Σ_n-elementary map $\bar{\pi} : J_\delta \to J_\beta$. We also use

FACT. For all $B \subset \varrho_M$, $B \in \Sigma_1(M')$ iff $B \in \Sigma_2(M)$.

Lemma B.2 gives the proof when M is admissible. It is, however, possible for M to be admissible and for M' not to be admissible.

For M which is $(n-1)$-sound, we let $\varrho^n_M = \varrho_{M(n-1)}$. The above facts show that ϱ^n_M is the least γ such that $\Sigma_n(M) \cap \mathcal{P}(\omega\gamma) \not\subset M$, the greatest γ such that for all $B \in \Sigma_n(M) \cap \mathcal{P}(\omega\gamma)$, $(J_{\varrho^n_M}, B)$ is amenable and that there is a partial function $p_n : \omega\varrho^n_M \twoheadrightarrow M$ which is $\Sigma_n(M)$. Combining with the canonical Σ_1–Skolem function, we get

COROLLARY C.5 (Jensen's uniformization theorem). *For all β, every $\Sigma_n(J_\beta)$ relation can be uniformized by a $\Sigma_n(J_\beta)$ function.*

APPENDIX D

The Admissible Collapse

In this appendix we review in a unified setting various results of Maass's ([**Maa78**]) and Sy Friedman's ([**Fri82**]) regarding isomorphisms between cones of Turing degrees and α-degrees. All references will refer to these two papers.

DEFINITION. An amenable $B \subset \alpha$ is *collapsible* if there is some bijection $p : \omega \iff \alpha$, weakly recursive in B. A *collapse* of such a degree is some $\mathcal{B} \subset \omega$ such that for all $A \subset \omega$, A is ω-r.e. in $\mathcal{B}$ iff it is α-r.e. in B.

First, we justify the names:

THEOREM D.1 (Maass). *Every collapsible α-degree with an amenable element has a collapse.*

PROOF. Let B be amenable and let $p : \omega \iff \alpha$ be a bijection, weakly recursive in B. Let

$$\mathcal{B} = \{(n, e, s) : p(n) \in \Phi(B)[p(s)], \text{ where } \Phi = p(e)\}.$$

$\mathcal{B}$ codes the canonical enumerations of all sets r.e. in B.

If A is ω-r.e. in $\mathcal{B}$, then B can enumerate it α-recursively, because B can α-enumerate all finite substrings of $\mathcal{B}$.

In the other direction, note that for every $C \subset \alpha$ which is α-r.e. in B, $p^{-1}C$ is ω-r.e. in $\mathcal{B}$. If $A \subset \omega$ is α-r.e. in B then $C = p``A$ is also α-r.e. in B which implies that $A = p^{-1}C$ is ω-r.e. in $\mathcal{B}$. □

Next we show some criteria sufficient for the existence of collapsible α-degrees. An amenable structure $M = (J_\beta, X)$ is *weakly admissible* if $\mathrm{cf}_{\Sigma_1(M)}(\omega\beta) \geqslant \varrho_M$.

LEMMA D.2. *Suppose that α is admissible and that $M = (J_\alpha, A)$ is weakly admissible. Then there is a bijection $f : \mathrm{cf}_{\Sigma_1(M)} \iff \alpha$ which is $\Sigma_1(M)$.*

PROOF. See [**Sac90**, IX 2.6]. □

THEOREM D.3 (Shore). *If A is amenable, r.e. and incomplete (i.e. $0' \nleqslant_\alpha A$) then*

$$\mathrm{rcf}(A) \geqslant \varrho_{\alpha,A},$$

that is, $(J_\alpha, \in, A)$ is weakly admissible.

PROOF. See [**Sho76b**]. □

Thus every amenable, incomplete r.e. set with recursive cofinality ω is collapsible.

Let B be collapsible, and let $\mathcal{B}$ be a collapse of B.

THEOREM D.4 (Friedmann). *Every (J_α, B)-degree contains an amenable set.*

In fact, since B is amenable, it follows that every α-degree above B contains an amenable set. Thus every α-degree above B contains a collapsible set.

LEMMA D.5. *The collapse is a well-defined embedding (preserving $\not\leqslant$) of the (J_α, B)-degrees into the $(V_\omega, \mathcal{B})$-degrees.*

PROOF. Suppose that $A, C \geqslant_\alpha B$ are amenable and let $\mathcal{A}, \mathcal{C}$ be any collapses of A, C respectively.

Suppose that $A \leqslant_\alpha C$. $\mathcal{A}$ and $\neg\mathcal{A}$ are ω-r.e. in $\mathcal{A}$, hence are α-r.e. in A, hence are α-r.e. in C, hence are ω-r.e. in $\mathcal{C}$, so $\mathcal{A} \leqslant_T \mathcal{C}$.

Suppose that $\mathcal{A} \leqslant_T \mathcal{C}$. Let $p : \omega \Longleftrightarrow \alpha$ be a bijection, weakly recursive in B. Consider $\hat{A} = \{K \in L_\alpha : K \subset A\}$ and $\widehat{\neg A} = \{K \in L_\alpha : K \subset \neg A\}$. Both sets are α-r.e. in A, hence $p^{-1}\hat{A}$ and $p^{-1}\widehat{\neg A}$ are α-r.e. in A, hence are ω-r.e. in $\mathcal{A}$, hence are ω-r.e. in $\mathcal{C}$, hence are α-r.e. in C, and so $\hat{A}$ and $\widehat{\neg A}$ are α-r.e. in C, so $A \leqslant_\alpha C$. □

In fact, we want to show that the collapse map is an isomorphism.

DEFINITION. A set $C \subset \omega$ is *immune* if every α-finite partial string $q \subset C$ with $\operatorname{dom} q \subset \omega$ is finite.

LEMMA D.6. *The following are equivalent for immune $A, C \subset \omega$:*

- $A \leqslant_T C \oplus \mathcal{B}$.
- $A \leqslant_{w\alpha} C \oplus B$.
- $A \leqslant_\alpha C \oplus B$.

PROOF. We are asking whether an enumeration of finite (=α-finite) substrings of A can be done with oracle C by a functional recursive in B, or in $\mathcal{B}$. Since we can restrict ourselves to ω, this is the same question. □

THEOREM D.7 (Maass, Friedman). *Every (J_α, B)-degree contains an immune set, and every $(V_\omega, \mathcal{B})$-degree contains an immune set.*

It follows that taking a B-degree of an immune set A to the $\mathcal{B}$-degree of A is well-defined and is an isomorphism between the B degrees and the $\mathcal{B}$ degrees. We want to show that this map is the same as the admissible collapse. We need to show

LEMMA D.8. *Let $A \geqslant_\alpha B$ be amenable, let $\mathcal{A}$ be its collapse, and let E be an immune set such that $E \equiv_B A$. Then $E \equiv_{\mathcal{B}} \mathcal{A}$.*

PROOF. $\mathcal{A} \leqslant_{w\alpha} A$, and so $\mathcal{A} \leqslant_{wB} E$. This implies that there is some weak (J_α, B)-functional Φ such that $\Phi(E) = \mathcal{A}$. Consider $\Phi_0 = \Phi \cap V_\omega$. Φ_0 is still α-r.e. in B, and $\Phi_0(E) = \mathcal{A}$ because E is immune. Φ_0 is ω-r.e. in $\mathcal{B}$, so $\mathcal{A} \leqslant_T \mathcal{B} \oplus E$, in other words, $\mathcal{A} \leqslant_{\mathcal{B}} E$.

On the other hand, $E \leqslant_{w\alpha} A$, and so $E \leqslant_T \mathcal{A}$. □

Let j be the inverse of the collapse map, from the Turing degrees above $\mathcal{B}$ to the α-degrees above B (equivalently, from the $\mathcal{B}$-degrees to the B-degrees).

The Collapse and Recursively Enumerable Degrees.

LEMMA D.9 (Maass, Friedman). *A $\mathcal{B}$-degree $\mathbf{d}$ contains a set ω-r.e. in $\mathcal{B}$ iff $j(\mathbf{d})$ contains a set which is strongly α-r.e. in B.*

COROLLARY D.10. *For every $\mathcal{B}$-degrees $\mathbf{d}_1 \leqslant \mathbf{d}_2$, $\mathbf{d}_2$ contains a set ω-r.e. in $\mathbf{d}_1$ iff $j(\mathbf{d}_2)$ contains a set strongly α-r.e. in $j(\mathbf{d}_1)$.*

PROOF. Lemma D.9 implies that a Turing degree $\mathbf{d}$ above $\mathcal{B}$ contains a set ω-r.e. in $\mathcal{B}$ iff $j(\mathbf{d})$ contains a set α-r.e. in B. Now relativize this lemma, using $\mathbf{d}_1$ as the base of the cone. □

THEOREM D.11 (Maass). *An α-degree $\mathbf{d}$ above B is r.e. in B iff $j^{-1}(\mathbf{d})$ is r.e. and above a degree $\mathbf{a}$ which is r.e. and above $\mathcal{B}$.*

As a corollary, we have that for every Turing degree $\mathbf{d}$ above $\mathcal{B}$, $j(\mathbf{d}'') = j(\mathbf{d})'$.

APPENDIX E

Prompt Permission

In this appendix we develop some of the theory of promptly simple sets in the context of admissible recursion theory. In fact, in the context of the r.e. degrees (rather than the lattice of sets), the basic notion seems to be that of prompt permitting.

DEFINITION. An amenable r.e. set A *permits promptly* if there is some effective enumeration $\langle A[s]\rangle$ of A and a recursive function p such that for all s, $p(s) \geqslant s$, and for all e such that W_e is unbounded in α, there is some $x \in W_e$ which enters W_e at some s (according to the canonical enumeration of the W_es) and such that

$$A \upharpoonright x+1\,[s] \neq A \upharpoonright x+1\,[p(s)].$$

LEMMA E.1 (Uniform Slowdown Lemma). *There is a recursive function f such that for all ϵ, if ϵ is a recursive index for an array of enumerations $\langle U_e[s]\rangle$ of r.e. sets U_e, then $f(\epsilon)$ is a recursive index for a total recursive function g such that for all e, $W_{g(e)} = U_e$ and every number in U_e enters U_e (according to the given approximation) before it enters $W_{g(e)}$ (according to the canonical enumeration).*

PROOF. We observe that the classical proof of the slowdown lemma (see [**Soa87**, XIII 1.5]) is easily generalized to the admissible context and is uniform. For any recursive index ϵ, let

$$U_e^{\epsilon}[s] \;=\; \cup\{\varphi_\epsilon(e,x) \,:\, x<s, \varphi_\epsilon(e,x)\downarrow[s]\}.$$

Find some recursive function h such that for all ϵ, e, i,

$$W_{h(\epsilon,e,i)} \;=\; \{x \,:\, \exists s\,(x \in U_e^{\epsilon}[s] \setminus W_i[s])\}.$$

By the recursion theorem with parameters, find some recursive function k such that for all e and ϵ, $W_{k(\epsilon,e)} \;=\; W_{h(\epsilon,e,k(\epsilon,e))}$. Let $f(\epsilon)$ be the index for $k(\epsilon,-)$. □

LEMMA E.2. *Suppose that $\langle U_e[s]\rangle$ is an effective enumeration of an array of r.e. sets U_e such that there is a recursive bound for $\cup\{U_e[s] \,:\, e \leqslant s\}$, and suppose that A permits promptly. Then there is some recursive function q such that for all s, $q(s) \geqslant s$ and such that for all e, if U_e is unbounded then there is some $x \in U_e$ which enters U_e at s and such that*

$$A \upharpoonright x+1\,[s] \neq A \upharpoonright x+1\,[q(s)].$$

An index for q can be obtained uniformly from an index for $\langle U_e[s]\rangle$.

PROOF. Suppose that p and $\langle a[s]\rangle$ witness that A permits promptly. Find a recursive g such that for all e, $W_{g(e)} = U_e$ and the former is enumerated more slowly. Define

$$q(s) \;=\; \sup\{p(t) \,:\, \exists x, e\,(x \text{ enters } U_e \text{ at } s \;\;\&\;\; x \text{ enters } W_{g(e)} \text{ at } t)\}.$$

An index for q can be obtained from an index for g. □

The last lemma justifies the technique used in chapter 3 to use prompt permission. In the verifications of that chapter, we use a variety of r.e. sets constructed during the construction and argue that during the construction, those sets which are unbounded received prompt permission. This is legal because by the recursion theorem, we could use a permitting function p which permits with respect to the sets being constructed.

Bibliography

[Cho84] C. T. Chong, *Techniques of admissible recursion theory*, Lecture Notes in Mathematics, vol. 1106, Springer-Verlag, Berlin, 1984. MR 87f:03001

[Chu38] Alonzo Church, *The constructive second number class*, Bull. Amer. Math. Soc. **44** (1938), 224–232.

[CK37] Alonzo Church and Stephen Cole Kleene, *Formal definitions in the theory of ordinal numbers*, Fund. Math. **28** (1937), 11–21.

[Dod82] A. J. Dodd, *The core model*, London Mathematical Society Lecture Note Series, vol. 61, Cambridge University Press, Cambridge, 1982. MR 84a:03062

[Dri68] Graham C. Driscoll, Jr., *Metarecursively enumerable sets and their metadegrees*, J. Symbolic Logic **33** (1968), 389–411. MR 43 #46

[DS96] Rod Downey and Richard A. Shore, *Lattice embeddings below a nonlow$_2$ recursively enumerable degree*, Israel J. Math. **94** (1996), 221–246. MR 97d:03056

[Fej82] Peter A. Fejer, *Branching degrees above low degrees*, Trans. Amer. Math. Soc. **273** (1982), no. 1, 157–180. MR 84a:03044

[Fri82] Sy D. Friedman, *The Turing degrees and the metadegrees have isomorphic cones*, Patras Logic Symposion (Patras, 1980), Stud. Logic Foundations Math., vol. 109, North-Holland, Amsterdam, 1982, pp. 145–157. MR 84g:03070

[G$\ddot{3}$9] Kurt Friedrich Gödel, *Consistency-proof for the generalized continuum-hypothsis*, Proc. Nat. Acad. Sci. USA **25** (1939), 220–224.

[GSS] Noam Greenberg, Richard A. Shore, and Theodore A. Slaman, *The theory of the metarecursively enumerable degrees*, Submitted.

[HS82] Leo Harrington and Saharon Shelah, *The undecidability of the recursively enumerable degrees*, Bull. Amer. Math. Soc. (N.S.) **6** (1982), no. 1, 79–80. MR 83i:03067

[Jec03] Thomas Jech, *Set theory*, Springer Monographs in Mathematics, Springer-Verlag, Berlin, 2003, The third millennium edition, revised and expanded. MR 2004g:03071

[Jen72] R. Björn Jensen, *The fine structure of the constructible hierarchy*, Ann. Math. Logic **4** (1972), 229–308; erratum, ibid. **4 (1972), 443**, With a section by Jack Silver. MR 46 #8834

[Kle38] Stephen Cole Kleene, *On notations for ordinal numbers*, J. Symbolic Logic **3** (1938), 150–155.

[Kre61] G. Kreisel, *Set theoretic problems suggested by the notion of potential totality.*, Infinitistic Methods (Proc. Sympos. Foundations of Math., Warsaw, 1959), Pergamon, Oxford, 1961, pp. 103–140. MR 26 #3599

[Kri64] Saul Kripke, *Transfinite recursion on admissible ordinals i, ii (abstracts)*, J. Symbolic Logic **29** (1964), 161–162.

[KS63] G. Kreisel and Gerald E. Sacks, *Metarecursive sets i, ii (abstracts)*, J. Symbolic Logic **28** (1963), 304–305.

[KS65] ———, *Metarecursive sets*, J. Symbolic Logic **30** (1965), 318–338. MR 35 #4097

[Lac72] A. H. Lachlan, *Embedding nondistributive lattices in the recursively enumerable degrees*, Conference in Mathematical Logic—London '70 (Proc. Conf., Bedford Coll., London, 1970), Springer, Berlin, 1972, pp. 149–177. Lecture Notes in Math., Vol. 255. MR 51 #12494

[Lac76] Alistair H. Lachlan, *A recursively enumerable degree which will not split over all lesser ones*, Ann. Math. Logic **9** (1976), no. 4, 307–365. MR 53 #12912

[Lac79] A. H. Lachlan, *Bounding minimal pairs*, J. Symbolic Logic **44** (1979), no. 4, 626–642. MR 80k:03041

[Ler74] Manuel Lerman, *Maximal α-r.e. sets*, Trans. Amer. Math. Soc. **188** (1974), 341–386. MR 48 #10785

[LS72] Manuel Lerman and Gerald E. Sacks, *Some minimal pairs of α-recursively enumerable degrees*, Ann. Math. Logic **4** (1972), 415–442. MR 55 #12491

[LS73] M. Lerman and S. G. Simpson, *Maximal sets in α-recursion theory*, Israel J. Math. **14** (1973), 236–247. MR 47 #8271

[Maa78] Wolfgang Maass, *Inadmissibility, tame r.e. sets and the admissible collapse*, Ann. Math. Logic **13** (1978), no. 2, 149–170. MR 80a:03060

[Myt89] Michael Mytilinaios, *Finite injury and Σ_1-induction*, J. Symbolic Logic **54** (1989), no. 1, 38–49. MR 90i:03067a

[NSS98] André Nies, Richard A. Shore, and Theodore A. Slaman, *Interpretability and definability in the recursively enumerable degrees*, Proc. London Math. Soc. (3) **77** (1998), no. 2, 241–291. MR 99m:03083

[Ode83] David Odell, *Trace constructions in α-recursion theory*, Ph.D. thesis, Cornell University, Ithaca, NY, 1983.

[Pla65] Richard Platek, *Foundations of recursion theory*, Ph.D. thesis, Stanford Univeristy, Stanford, CA, 1965.

[Sac63] Gerald E. Sacks, *Degrees of unsolvability*, Princeton University Press, Princeton, N.J., 1963. MR 32 #4013

[Sac66] ———, *Metarecursively enumerable sets and admissible ordinals*, Bull. Amer. Math. Soc. **72** (1966), 59–64. MR 35 #6542

[Sac90] ———, *Higher recursion theory*, Perspectives in Mathematical Logic, Springer-Verlag, Berlin, 1990. MR 92a:03062

[Sho76a] Richard A. Shore, *On the jump of an α-recursively enumerable set*, Trans. Amer. Math. Soc. **217** (1976), 351–363. MR 54 #12504

[Sho76b] ———, *The recursively enumerable α-degrees are dense*, Ann. Math. Logic **9** (1976), no. 1-2, 123–155. MR 52 #2852

[Sho78] ———, *On the $\forall\exists$-sentences of α-recursion theory*, Generalized recursion theory, II (Proc. Second Sympos., Univ. Oslo, Oslo, 1977), Stud. Logic Foundations Math., vol. 94, North-Holland, Amsterdam, 1978, pp. 331–353. MR 80e:03055

[Sho82] ———, *On homogeneity and definability in the first-order theory of the Turing degrees*, J. Symbolic Logic **47** (1982), no. 1, 8–16. MR 84a:03046

[Sho97] ———, *Conjectures and questions from Gerald Sacks's degrees of unsolvability*, Arch. Math. Logic **36** (1997), no. 4-5, 233–253, Sacks Symposium (Cambridge, MA, 1993). MR 99a:03043

[Sho99] ———, *The recursively enumerable degrees*, Handbook of computability theory, Stud. Logic Found. Math., vol. 140, North-Holland, Amsterdam, 1999, pp. 169–197. MR 2000j:03057

[Sla91] Theodore A. Slaman, *Degree structures*, Proceedings of the International Congress of Mathematicians, Vol. I, II (Kyoto, 1990) (Tokyo), Math. Soc. Japan, 1991, pp. 303–316. MR 93b:03074

[Soa87] Robert I. Soare, *Recursively enumerable sets and degrees*, Perspectives in Mathematical Logic, Springer-Verlag, Berlin, 1987, A study of computable functions and computably generated sets. MR 88m:03003

[SS72] G. E. Sacks and S. G. Simpson, *The α-finite injury method*, Ann. Math. Logic **4** (1972), 343–367. MR 51 #5277

[SS90] Juichi Shinoda and Theodore A. Slaman, *On the theory of the PTIME degrees of the recursive sets*, J. Comput. System Sci. **41** (1990), no. 3, 321–366. MR 92b:03049

[Suk69] J. Sukonick, *Lower bounds for pairs of metarecursively enumerable degrees*, Ph.D. thesis, M.I.T., Cambridge, MA, 1969.

[SW89] Theodore A. Slaman and W. Hugh Woodin, *Σ_1-collection and the finite injury priority method*, Mathematical logic and applications (Kyoto, 1987), Lecture Notes in Math., vol. 1388, Springer, Berlin, 1989, pp. 178–188. MR 91j:03075

[Tak60] Gaisi Takeuti, *On the recursive functions of ordinal numbers*, J. Math. Soc. Japan **12** (1960), 119–128. MR 23 #A1524

[Tak65] ———, *A formalization of the theory of ordinal numbers*, J. Symbolic Logic **30** (1965), 295–317. MR 33 #5467

[Wei88] B. Weinstein, *On embeddings of the 1-3-1 lattice into the recursively enumerable degrees*, Ph.D. thesis, University of California, Berkeley, 1988.

Editorial Information

To be published in the *Memoirs*, a paper must be correct, new, nontrivial, and significant. Further, it must be well written and of interest to a substantial number of mathematicians. Piecemeal results, such as an inconclusive step toward an unproved major theorem or a minor variation on a known result, are in general not acceptable for publication. Papers appearing in *Memoirs* are generally at least 80 and not more than 200 published pages in length. Papers less than 80 or more than 200 published pages require the approval of the Managing Editor of the Transactions/Memoirs Editorial Board.

As of January 31, 2006, the backlog for this journal was approximately 14 volumes. This estimate is the result of dividing the number of manuscripts for this journal in the Providence office that have not yet gone to the printer on the above date by the average number of monographs per volume over the previous twelve months, reduced by the number of volumes published in four months (the time necessary for preparing a volume for the printer). (There are 6 volumes per year, each containing at least 4 numbers.)

A Consent to Publish and Copyright Agreement is required before a paper will be published in the *Memoirs*. After a paper is accepted for publication, the Providence office will send a Consent to Publish and Copyright Agreement to all authors of the paper. By submitting a paper to the *Memoirs*, authors certify that the results have not been submitted to nor are they under consideration for publication by another journal, conference proceedings, or similar publication.

Information for Authors

Memoirs are printed from camera copy fully prepared by the author. This means that the finished book will look exactly like the copy submitted.

The paper must contain a *descriptive title* and an *abstract* that summarizes the article in language suitable for workers in the general field (algebra, analysis, etc.). The *descriptive title* should be short, but informative; useless or vague phrases such as "some remarks about" or "concerning" should be avoided. The *abstract* should be at least one complete sentence, and at most 300 words. Included with the footnotes to the paper should be the 2000 *Mathematics Subject Classification* representing the primary and secondary subjects of the article. The classifications are accessible from `www.ams.org/msc/`. The list of classifications is also available in print starting with the 1999 annual index of *Mathematical Reviews*. The Mathematics Subject Classification footnote may be followed by a list of *key words and phrases* describing the subject matter of the article and taken from it. Journal abbreviations used in bibliographies are listed in the latest *Mathematical Reviews* annual index. The series abbreviations are also accessible from `www.ams.org/publications/`. To help in preparing and verifying references, the AMS offers MR Lookup, a Reference Tool for Linking, at `www.ams.org/mrlookup/`. When the manuscript is submitted, authors should supply the editor with electronic addresses if available. These will be printed after the postal address at the end of the article.

Electronically prepared manuscripts. The AMS encourages electronically prepared manuscripts, with a strong preference for $\mathcal{A}_\mathcal{M}\mathcal{S}$-LaTeX. To this end, the Society has prepared $\mathcal{A}_\mathcal{M}\mathcal{S}$-LaTeX author packages for each AMS publication. Author packages include instructions for preparing electronic manuscripts, the *AMS Author Handbook*, samples, and a style file that generates the particular design specifications of that publication series. Though $\mathcal{A}_\mathcal{M}\mathcal{S}$-LaTeX is the highly preferred format of TeX, author packages are also available in $\mathcal{A}_\mathcal{M}\mathcal{S}$-TeX.

Authors may retrieve an author package from e-MATH starting from `www.ams.org/tex/` or via FTP to `ftp.ams.org` (login as `anonymous`, enter username as password, and type `cd pub/author-info`). The *AMS Author Handbook* and the *Instruction Manual* are available in PDF format following the author packages link from `www.ams.org/tex/`. The author package can also be obtained free of charge by sending email to tech-support@ams.org (Internet) or from the Publication Division, American Mathematical Society, 201 Charles St., Providence, RI 02904-2294, USA. When requesting an author package, please specify $\mathcal{A}\mathcal{M}\mathcal{S}$-LaTeX or $\mathcal{A}\mathcal{M}\mathcal{S}$-TeX and the publication in which your paper will appear. Please be sure to include your complete mailing address.

Sending electronic files. After acceptance, the source file(s) should be sent to the Providence office (this includes any TeX source file, any graphics files, and the DVI or PostScript file).

Before sending the source file, be sure you have proofread your paper carefully. The files you send must be the EXACT files used to generate the proof copy that was accepted for publication. For all publications, authors are required to send a printed copy of their paper, which exactly matches the copy approved for publication, along with any graphics that will appear in the paper.

TeX files may be submitted by email, FTP, or on diskette. The DVI file(s) and PostScript files should be submitted only by FTP or on diskette unless they are encoded properly to submit through email. (DVI files are binary and PostScript files tend to be very large.)

Electronically prepared manuscripts can be sent via email to pub-submit@ams.org (Internet). The subject line of the message should include the publication code to identify it as a Memoir. TeX source files, DVI files, and PostScript files can be transferred over the Internet by FTP to the Internet node e-math.ams.org (130.44.1.100).

Electronic graphics. Comprehensive instructions on preparing graphics are available at www.ams.org/jourhtml/graphics.html. A few of the major requirements are given here.

Submit files for graphics as EPS (Encapsulated PostScript) files. This includes graphics originated via a graphics application as well as scanned photographs or other computer-generated images. If this is not possible, TIFF files are acceptable as long as they can be opened in Adobe Photoshop or Illustrator. No matter what method was used to produce the graphic, it is necessary to provide a paper copy to the AMS.

Authors using graphics packages for the creation of electronic art should also avoid the use of any lines thinner than 0.5 points in width. Many graphics packages allow the user to specify a "hairline" for a very thin line. Hairlines often look acceptable when proofed on a typical laser printer. However, when produced on a high-resolution laser imagesetter, hairlines become nearly invisible and will be lost entirely in the final printing process.

Screens should be set to values between 15% and 85%. Screens which fall outside of this range are too light or too dark to print correctly. Variations of screens within a graphic should be no less than 10%.

Inquiries. Any inquiries concerning a paper that has been accepted for publication should be sent directly to the Electronic Prepress Department, American Mathematical Society, 201 Charles St., Providence, RI 02904, USA.

Editors

This journal is designed particularly for long research papers, normally at least 80 pages in length, and groups of cognate papers in pure and applied mathematics. Papers intended for publication in the *Memoirs* should be addressed to one of the following editors. In principle the Memoirs welcomes electronic submissions, and some of the editors, those whose names appear below with an asterisk (*), have indicated that they prefer them. However, editors reserve the right to request hard copies after papers have been submitted electronically. Authors are advised to make preliminary email inquiries to editors about whether they are likely to be able to handle submissions in a particular electronic form.

***Algebra** to ALEXANDER KLESHCHEV, Department of Mathematics, University of Oregon, Eugene, OR 97403-1222; email: ams@noether.uoregon.edu

Algebra and its application to MINA TEICHER, Emmy Noether Research Institute for Mathematics, Bar-Ilan University, Ramat-Gan 52900, Israel; email: teicher@macs.biu.ac.il

Algebraic geometry to DAN ABRAMOVICH, Department of Mathematics, Brown University, Box 1917, Providence, RI 02912; email: amsedit@math.brown.edu

***Algebraic number theory** to V. KUMAR MURTY, Department of Mathematics, University of Toronto, 100 St. George Street, Toronto, ON M5S 1A1, Canada; email: murty@math.toronto.edu

***Algebraic topology** to ALEJANDRO ADEM, Department of Mathematics, University of British Columbia, Room 121, 1984 Mathematics Road, Vancouver, British Columbia, Canada V6T 1Z2; email: adem@math.ubc.ca

Combinatorics to JOHN R. STEMBRIDGE, Department of Mathematics, University of Michigan, Ann Arbor, Michigan 48109-1109; email: FRS@umich.edu

Complex analysis and harmonic analysis to ALEXANDER NAGEL, Department of Mathematics, University of Wisconsin, 480 Lincoln Drive, Madison, WI 53706-1313; email: nagel@math.wisc.edu

***Differential geometry and global analysis** to LISA C. JEFFREY, Department of Mathematics, University of Toronto, 100 St. George St., Toronto, ON Canada M5S 3G3; email: jeffrey@math.toronto.edu

Dynamical systems and ergodic theory to AMIE WILKINSON, Department of Mathematics, Northwestern University, 2033 Sheridan Road, Evanston, IL 60208-2730; email: wilkinso@math.northwestern.edu

***Functional analysis and operator algebras** to MARIUS DADARLAT, Department of Mathematics, Purdue University, 150 N. University St., West Lafayette, IN 47907-2067; email: mdd@math.purdue.edu

***Geometric analysis** to TOBIAS COLDING, Courant Institute, New York University, 251 Mercer St., New York, NY 10012; email: traneditor@cims.nyu.edu

***Geometric analysis** to MLADEN BESTVINA, Department of Mathematics, University of Utah, 155 South 1400 East, JWB 233, Salt Lake City, Utah 84112-0090; email: bestvina@math.utah.edu

Harmonic analysis, representation theory, and Lie theory to ROBERT J. STANTON, Department of Mathematics, The Ohio State University, 231 West 18th Avenue, Columbus, OH 43210-1174; email: stanton@math.ohio-state.edu

***Logic** to STEFFEN LEMPP, Department of Mathematics, University of Wisconsin, 480 Lincoln Drive, Madison, Wisconsin 53706-1388; email: lempp@math.wisc.edu

***Ordinary differential equations, and applied mathematics** to PETER W. BATES, Department of Mathematics, Michigan State University, East Lansing, MI 48824-1027; email: bates@math.msu.edu

***Partial differential equations** to GUSTAVO PONCE, Department of Mathematics, South Hall, Room 6607, University of California, Santa Barbara, CA 93106; email: ponce@math.ucsb.edu

***Probability and statistics** to KRZYSZTOF BURDZY, Department of Mathematics, University of Washington, Box 354350, Seattle, Washington 98195-4350; email: burdzy@math.washington.edu

***Real analysis and partial differential equations** to DANIEL TATARU, Department of Mathematics, University of California, Berkeley, Berkeley, CA 94720; email: tataru@math.berkeley.edu

All other communications to the editors should be addressed to the Managing Editor, ROBERT GURALNICK, Department of Mathematics, University of Southern California, Los Angeles, CA 90089-1113; email: guralnic@math.usc.edu.

Titles in This Series

856 **Vladimir Bolotnikov and Harry Dym,** On boundary interpolation for matrix valued Schur functions, 2006

855 **Yevgenia Kashina, Yorck Sommerhäuser, and Yongchang Zhu,** On higher Frobenius-Schur indicators, 2006

854 **Noam Greenberg,** The role of true finiteness in the admissible recursively enumerable degrees, 2006

853 **Joachim Krieger,** Stability of spherically symmetric wave maps, 2006

852 **Viorel Barbu, Irena Lasiecka, and Roberto Triggiani,** Tangential boundary stabilization of Navier-Stokes equations, 2006

851 **Jie Wu,** On maps from loop suspensions to loop spaces and the shuffle relations on the Cohen groups, 2006

850 **Siegfried Echterhoff, S. Kaliszewski, John Quigg, and Iain Raeburn,** A categorical approach to imprimitivity theorems for C^*-dynamical systems, 2006

849 **Katsuhiko Kuribayashi, Mamoru Mimura, and Tetsu Nishimoto,** Twisted tensor products related to the cohomology of the classifying spaces of loop groups, 2006

848 **Bob Oliver,** Equivalences of classifying spaces completed at the prime two, 2006

847 **Eric T. Sawyer and Richard L. Wheeden,** Hölder continuity of weak solutions to subelliptic equations with rough coefficients, 2006

846 **Victor Beresnevich, Detta Dickinson, and Sanju Velani,** Measure theoretic laws for lim–sup sets, 2006

845 **Ehud Friedgut, Vojtech Rödl, Andrzej Ruciński, and Prasad V. Tetali,** A Sharp threshold for random graphs with a monochromatic triangle in every edge coloring, 2006

844 **Amadeu Delshams, Rafael de la Llave, and Tere M. Seara,** A geometric mechanism for diffusion in Hamiltonian systems overcoming the large gap problem: Heuristics and rigorous verification on a model, 2006

843 **Denis V. Osin,** Relatively hyperbolic groups: Intrinsic geometry, algebraic properties, and algorithmic problems, 2006

842 **David P. Blecher and Vrej Zarikian,** The calculus of one-sided M-ideals and multipliers in operator spaces, 2006

841 **Enrique Artal Bartolo, Pierrette Cassou-Noguès, Ignacio Luengo, and Alejandro Melle Hernández,** Quasi-ordinary power series and their zeta functions, 2005

840 **Sławomir Kołodziej,** The complex Monge-Ampère equation and pluripotential theory, 2005

839 **Mihai Ciucu,** A random tiling model for two dimensional electrostatics, 2005

838 **V. Jurdjevic,** Integrable Hamiltonian systems on complex Lie groups, 2005

837 **Joseph A. Ball and Victor Vinnikov,** Lax-Phillips scattering and conservative linear systems: A Cuntz-algebra multidimensional setting, 2005

836 **H. G. Dales and A. T.-M. Lau,** The second duals of Beurling algbras, 2005

835 **Kiyoshi Igusa,** Higher complex torsion and the framing principle, 2005

834 **Kenίchi Ohshika,** Kleinian groups which are limits of geometrically finite groups, 2005

833 **Greg Hjorth and Alexander S. Kechris,** Rigidity theorems for actions of product groups and countable Borel equivalence relations, 2005

832 **Lee Klingler and Lawrence S. Levy,** Representation type of commutative Noetherian rings III: Global wildness and tameness, 2005

831 **K. R. Goodearl and F. Wehrung,** The complete dimension theory of partially ordered systems with equivalence and orthogonality, 2005

830 **Jason Fulman, Peter M. Neumann, and Cheryl E. Praeger,** A generating function approach to the enumeration of matrices in classical groups over finite fields, 2005

829 **S. G. Bobkov and B. Zegarlinski,** Entropy bounds and isoperimetry, 2005